Markers for Neural and Endocrine Cells

Edited by
M. Gratzl and K. Langley

© VCH Verlagsgesellschaft mbH, D-6940 Weinheim (Federal Republic of Germany), 1991

Distribution:

VCH, P.O. Box 10 11 61, D-6940 Weinheim (Federal Republic of Germany)

Switzerland: VCH, P.O. Box, CH-4020 Basel (Switzerland)

United Kingdom and Ireland: VCH (UK) Ltd., 8 Wellington Court, Cambridge CB 1 1HZ (England)

USA and Canada: VCH, Suite 909, 220 East 23rd Street, New York, NY 10010-4606 (USA)

ISBN 3-527-28167-3 (VCH, Weinheim)   ISBN 0-89573-986-0 (VCH, New York)

Markers for Neural and Endocrine Cells

Molecular and Cell Biology, Diagnostic Applications

Edited by
Manfred Gratzl and Keith Langley

VCH   Weinheim · New York · Basel · Cambridge

Editors:
Prof. Dr. Manfred Gratzl
Abteilung Anatomie und Zellbiologie
Universität Ulm
Albert-Einstein-Allee 11
D-7900 Ulm
Germany

Dr. Keith Langley
Unité INSERM U-44
Centre de Neurochimie
5, rue Blaise Pascal
F-67084 Strasbourg
France

Published jointly by
VCH Verlagsgesellschaft mbH, Weinheim (Federal Republic of Germany)
VCH Publishers, Inc., New York, NY (USA)

Editorial Director: Dr. Hans-Joachim Kraus
Production Manager: Max Denk

Cover illustration: Human small cell lung carcinoma cells immunostained for neural cell adhesion molecules (Photo: Dr. Keith Langley).

Library of Congress Card No. 90-13053.

British Library Cataloguing in Publication Data
Markers for neural and endocrine cells.
  1. Medicine. Diagnosis. Use of cell makers
  I. Gratzl, M. (Manfred) II. Langley, Keith
  616.97582

  ISBN 3-527-28167-3 W. Germany
  ISBN 0-89673-986-0 US

Deutsche Bibliothek Cataloguing-in-Publication Data:

Markers for neural and endocrine cells : molecular and cell
biology, diagnostic applications / ed. by M. Gratzl and K.
Langley. – Weinheim ; New York ; Basel ; Cambridge : VCH,
1990
  ISBN 3-527-28167-3 (Weinheim ...)
  ISBN 0-89573-986-0 (New York)
NE: Gratzl, Manfred [Hrsg.]

© VCH Verlagsgesellschaft mbH, D-6940 Weinheim (Federal Republic of Germany), 1991

Printed on acid-free paper.

Composition: Mitterweger Werksatz GmbH, D-6831 Plankstadt, Brauereistr. 13 Printing: Zechner GmbH & Co. KG, D-6720 Speyer Bookbinding: Josef Spinner, Großbuchbinderei GmbH D-7583 Ottersweier
Printed in the Federal Republic of Germany

For Karin and Jean

Preface

The initial idea to compile a book on marker substances useful for characterising neurons and endocrine cells was conceived at a meeting held in the autumn of 1988 at Gargellen in the Austrian Alps under the auspices of the "Gesellschaft für Histochemie" (International Association of Histochemists). Informal discussions that took place between fundamental research workers and clinicians, both in that beautiful and isolated congress location and subsequently at other meetings and through personal contact pointed out the value of celltype markers in clinical practice and the constant need to test potentially new markers. They also made us aware of the need for an authoritative volume on the one hand reviewing the latest data on the molecular biology, biochemistry and cell biology of these marker molecules and on the other explaining how these molecules are employed clinically. This book represents the fruits of these discussions. In it we attempt to bridge the gap between fundamental research and clinical practice and hope that it will help to stimulate future dialogue between the basic disciplines and applied research.

The clinical aspects on which we chose to concentrate in this book principally concern abnormalities of the nervous and endocrine systems. Present day concepts on the relationships existing between these two systems owe a enormous debt to the eminent pathologist E. Feyrter and his pioneering work early this century, which culminated in his book "Über diffuse endokrine epitheliale Organe" published in Leipzig in 1938, and also to the renowned histochemists and cytochemists A. G. E. Pearse and T. Fujita in the 1960s and 1970s. Developmental aspects of neural and endocrine cells have since been studied in depth by Nicole Le Douarin and her colleagues and constitute the theme of the first chapter in this book.

While neurons and endocrine cells share many properties they also differ in important respects. The most striking differences include their basic morphology: neurons extend long processes (axons and dentrites) in contrast to the absence of such structures in endocrine cells. Neurons are also remarkable for their capacity to separate sites of synthesis of neurotransmitters from their sites of storage and release, while such separation is spatially very limited in endocrine cells. Neurotransmitter release occurs from neurons at specialized junctions (synapses, neuromuscular junctions) while endocrine cells secrete their hormones into the extracellular space, from where they may be transported in the blood to affect remote targets in the body. Exceptions to these general principles exist. Thus the hypothalamic/neurohypophyseal system bridges the gap between the neural and endocrine systems in that their cell bodies are part of the central nervous system while their distant endings release hormones in the vicinity of blood vessels. Other examples which do not happily fit into these broad divisions may be found (e.g. endocrine cells which affect their immediate neighbors in a paracrine fashion).

Apart from the developmental relationships reviewed in the first chapter, neurons and endocrine cells share many common proteins which may in part constitute elements of the secretory machinery. Chapters 2 and 3 concentrate on clear synaptic vesicles and secretory granules, respectively and focus attention in particular on either membrane constituents or proteins contained within granules. A class of molecules which profoundly affects both developmental events and pathological processes, the NCAM family, dealt with in chapter 4, is representative of cell surface membrane constituents common to neural and endocrine cells.

One of the first marker proteins used for identifying neurons and endocrine cells, the cytosolic enzyme neuron specific enolase, NSE, was also one of the first to attain the status of a clinically useful diagnostic tool (chapter 5). Secretory proteins contained in granules, such as chromogranin A, CGA, which are released from actively secreting cells are also now employed in clinical practice (chapter 6). Synaptophysin is a recent addition to the clinicians' battery of marker substances reviewed in chapter 7, which also discusses the value of general cell markers (including intermediate filament proteins) and specific peptide hormone markers. Different forms of NCAM, including post-translationally modified forms represent examples of molecules the potential applications of which have yet to be fully explored, though promising indications are already published. Other adhesion molecules as yet unexplored in this context may also share this potential.

If this book stimulates future contact between those who study the molecular biology, biochemistry or cell biology of given molecules and those who are seeking for better ways of diagnosing the condition of their patients, we will have achieved one of our principal aims. Without the aid and cooperation of the experts who contributed the different chapters, this would not have been possible.

July 1990,
Strasbourg and Ulm   Keith Langley and Manfred Gratzl

Table of Contents

**4   Neural Cell Adhesion Molecule NCAM in Neural and Endo-
            crine Cells**

Contributors

Dr. Pietro De Camilli
Department of Cell Biology
Yale University
333 Cedar Street
New Haven, CT 06510, USA

Prof. Nicole M. Le Douarin
Institut d'Embryologie
Cellulaire et Moléculaire
49, avenue de la Belle
Gabrielle
F-94736 Nogent-sur-Marne, France

Dr. Josiane Fontaine-Pérus
Faculté de Science et de Techniques
Université de Nantes
2 rue de la Houssinière
F-44072 Nantes, France

Prof. Dr. Manfred Gratzl
Abteilung Anatomie und
Zellbiologie der
Universität Ulm
Albert-Einstein-Allee 11
D-7900 Ulm, Germany

Prof. Dr. Philipp U. Heitz
Dr. Paul Komminoth
Dr. Christian Zuber
Institut für Pathologie
der Universität
Schmelzbergstraße 12
CH-8091 Zürich, Switzerland

Dr. Wieland B. Huttner
Dr. Hans-Hermann Gerdes
European Molecular Biology Lab.
Meyerhofstraße 1
D-6900 Heidelberg, Germany

Dr. Reinhard Jahn
Max-Planck-Institut
für Psychiatrie
Abt. Neurochemie
Am Klopferspitz 18A
D-8033 Martinsried, Germany

Dr. Keith Langley
Unité INSERM U-44
Centre de Neurochimie
5, rue Blaise Pascal
F-67084 Strasbourg, France

Dr. Paul J. Marangos
Gensia Pharmaceuticals, Inc.
11075 Roselle Street
San Diego, CA 92121-1207, USA

Daniel T. O'Connor, M.D.
Marwan A. Takiyyuddin, M.D.
Juan A. Barbosa, M.D.
Ray J. Hsiao, M.D.
Robert J. Parmer, M.D.
Department of Medicine
Nephrology/Hypertension Division
V. A. Medical Center
3350 La Jolla Village Drive
San Diego, CA 92161, USA

Dr. Patrizia Rosa
Dipartimento di Farmacologia
Università di Milano
Via Vanvitelli 32
I-20129 Milano, Italy

Prof. Dr. Jürgen Roth
Biozentrum
Klingelbergstraße 70
CH-4056 Basel, Switzerland

Part I

Ontogenetic Relationships

1 Embryonic Origin of Polypeptide Hormone Producing Cells

Nicole M. Le Douarin and *Josiane Fontaine-Pérus*

1.1 Introduction

Since the discovery that all cells of the peripheral ganglia and nerves originate from the neural crest, the dogma that the nervous system of vertebrates is entirely of ectodermal origin has been universally accepted (see Le Douarin, 1982 for a review). It has long been known that the ectodermal germ layer produces the endocrine cells of the adenohypophysis through the Rathke's pouch and that the adrenergic autonomic neurons, the small intensely fluorescent (SIF) cells of the sympathetic ganglia and the chromaffin endocrine cells of the suprarenal gland and various adrenergic paraganglia belong to a common cell lineage (Landis and Patterson, 1981). It is therefore obvious that close developmental and functional relationships exist between neurons, either displaying or not a neuroendocrine activity, and certain endocrine cells, essentially those responsible for polypeptide hormone production.

Endocrinology and one of its clinical cornerstones, diabetology, originated from the interest that certain renowned physiologists of the nineteenth century took in the pancreas. It was by examining the postulate of Claude Bernard that this organ is necessary for survival (C. Bernard, 1856), that Mering and Minkowski (1889) discovered that *diabetes mellitus* was due to a disfunction of the pancreas. The concept of "blood borne chemical messengers" was thereafter proposed by Bayliss and Starling (1902) with the discovery of secretin, the first of what turned out to be a long list of hormones produced by the gut epithelium.

Although endocrinology began with the discovery of a hormone produced by the digestive epithelium, it is only considerably later that the gastrointestinal tract, became fully recognized as an endocrine organ. In 1938 Friedrich Feyrter described a system of clear cells (helle Zellen) dispersed in the gut epithelium and in various other parts of the body. He considered that at least some of these cells were acting on their immediate neighbours and were therefore paracrine in nature, rather than endocrine. He thought that they arose from the enterocytes by a process called *endophytie* and are therefore of endodermal origin.

The fact that gut endocrinology progressed so slowly and that "most gut hormones rested unknown in the darkness of the bowel"[1] is due to several reasons, mainly because they are not concentrated in glands like steroid, or thyroid and pituitary hormones for example, the knowledge of which progressed so rapidly during the first half of this century. Such a dispersion of the secretory cells hampered the collection of materials for chemical purification and physiological investigation.

However, their identification has dramatically progressed since the sixties, when efficient purification procedures were used to isolate and then sequence several polypeptidic hormones from the digestive tract. The first of these followed by many others, were mammalian gastrins by Gregory and Tracy (1964) and secretin by Viktor Mutt and his collaborators (1970).

At the same time the embryonic origin of the "system of clear cells" as described by Feyrter was questioned and the initial conception of a diffuse endocrine organ was reinvestigated and developed with new experimental methods by A.G.E. Pearse.

Pearse emphasized the fact that a number of polypeptide-hormone-producing cells, located mainly in the gut and its appendages but also in other parts of the body, possess a common set of cytochemical and ultrastructural characteristics (Pearse, 1968, 1969) Tab. 1-1.

Tab. 1-1 Cytochemical characteristics of polypeptide-hormone-secreting cells of the APUD series (from Pearse, 1969).

(A) 1. Fluorogenic **a**mine content (catecholamines, 5 HT or others)
      (a) primary; (b), secondary uptake
(P) 2. Amine **p**recursor **u**ptake (5-HTP, DOPA)
(U)
(D) 3. Amino acid **d**ecarboxylase
    4. High side-chain carboxyl or carboxyamide (masked metachromasia)
    5. High non-specific esterase or cholinesterase (or both)
    6. High α-GPD (α-glycerophosphate menadione reductase)
    7. Specific immunofluorescence

5-HT, 5-hydroxytryptamine; 5-HTP, hydroxytryptophan; DOPA, 3,4-dihydroxyphenylalanine; α-GPD, α-glycerophosphate dehydrogenase

---

[1] quoted from Rehfeld, J.F. (1981) In: *Gut Hormones*, (S.R. Bloom & J.M. Polak, eds.), Churchill Livingstone, p. 10.

Pearse grouped these cells in the so-called APUD series, an acronym derived from the initial letters of their three more constant and important cytochemical properties: **A**mine content and/or **A**mine **P**recursor **U**ptake and **D**ecarboxylation.

For Pearse, these features indicated closely related metabolic mechanisms and common synthetic storage and secretion properties. On that account he proposed that the APUD cells share a common embryological precursor which arises from the neural crest. The list of cells belonging to the APUD series, limited to 14 cell types (including pituitary corticotrophs and melanotrophs, pancreatic islet cells, calcitonin producing cells, carotid body type I cells, adrenal medulla and various endocrine cells of the gut epithelium) in 1969, had increased to 40 less than 10 years later. By then Pearse had included the parathyroid and a number of cells in the gut, lung and skin that had been shown to produce peptides or neuropeptides (Pearse, 1976, 1977).

In all these cells the most characteristic properties can be demonstrated by a simple cytochemical test. The L-isomer of either of the two principal aminoacid precursors of the fluorogenic monoamines (3,4-dihydroxyphenylalanine (DOPA) for catecholamines or 5-hydroxytrypoptophan (5-HTP) for serotonin) administered intravenously is taken up and decarboxylated by APUD cells.

Some ultrastructural characteristics are shared by the APUD cells (Pearse, 1968):

1. low levels of rough (granular) endoplasmic reticulum;
2. high levels of smooth endoplasmic reticulum in the form of vesicles;
3. high content of free ribosomes;
4. electron-dense, fixation-labile mitochondria;
5. membrane-bound secretion vesicles with osmiophilic contents (average diameter 100–200 nm).

The first test case for the APUD concept was the cell type reported by Pearse (1968) responsible for the production of the calcium-regulating hormone, calcitonin, a hormone discovered by Copp and his associates (Copp et al., 1962). The source of calcitonin in the thyroid gland was identified as the parafollicular cells which were then renamed C-cells (C for calcitonin).

Finally, Pearse's theory was that cells of the APUD series constituted a "diffuse neuroendocrine system" (DNES) which he viewed as a third branch of the nervous system, "acting with the second, autonomic division, in the control of the function of all the intestinal organs". (Pearse, 1980).

One of the implications of the APUD cell concept was that a common origin for a variety of endocrine cells could account for a number of associated endocrine disorders such as medullary thyroid carcinomas concomitant with the pheochromocytome syndrome (Ljungberg et al., 1967), the Zollinger-Ellison syndrome (Zollinger and Ellison, 1955) and other forms of the so-called multiple endocrine tumors. This explains why, although based only on very circumstancial evidence, the APUD cell concept was widely accepted.

The neuroectodermal origin of the endocrine cells of the gut epithelium and of the endocrine glands (pancreatic islet cells) associated with the digestive tract is, however, a very controversial subject. On the one hand an impressive number of molecular markers have been shown to be shared by neurons and a variety of these polypeptide hormone producing cells, thus pleading for the existence of a common embryological ancestor, from which, in a terminal differentiation state, several cell types would have emerged. On the other hand, precise cell tracing techniques applied on the avian embryo, an experimental model particularly favourable for this type of analysis, have failed to confirm the implication of ectodermal cells in histogenesis of the gut related endocrine organs. These two sets of data will be briefly analyzed in this chapter from a developmental view point.

1.2 The Use of Molecular Markers to Study Cell Lineages

It is often assumed that when different cell types diverge from a common progenitor cell, certain metabolic characteristics are conserved and therefore can be considered as markers of their kinship. This may or may not be true and should be regarded only as an indication of a possible common origin essentially because when carefully examined, most of the markers do not show the strict cell-type specificity that one should expect in such a case.

For example, the most characteristic common property of APUD cells, i.e. decarboxylation of the aminoacid precursors of fluorogenic monoamines (DOPA and 5HTP) is due to the presence in these cells of an L-aminoacid decarboxylase (AADC). This enzyme is present in the neuroepithelium of the basal plate of brain and spinal cord of young rat embryos. However, during development, it is not restricted to neural and endocrine cells since it is also transiently expressed by other tissues of mesodermal origin such as notochord and muscles as well as in pancreatic acinar cells, the endodermic origin of which is universally accepted (Teitelman et al., 1981).

Another proposed marker for APUD cells is a cholinesterase. In chick embryos, a specific acetylcholinesterase (AChE) is found in developing pancreatic endocrine cells, in some unidentified cells in the gut groove (Drews et al., 1967), in the neuroepithelium of the basal plate of the neural tube, and in the neural crest (Cochard and Coltey, 1983). At later stages, however, AChE does not remain confined to neural and neuroendocrine cells; thus its status as a molecular marker for the DNES is doubtful.

Neuron-specific enolase (NSE), an enzyme present in nerve cells (Marangos et al., 1978), is also found in some cells of the diffuse endocrine system including entero-chromaffin cells and the pancreatic islet cells (Schmechel et al., 1978; Falkmer et al., 1984). However, NSE has been detected in other cell types, such as megakaryocytes and platelets (Marangos et al., 1980).

There are however other molecular features strikingly common to neurons and endocrine cells. This is the case for synaptophysin, an integral membrane glycoprotein localized in presynaptic vesicles of neurons and in similar vesicles of the adrenal medulla. It was also reported to be present in pancreatic islet cells and in a variety of epithelial tumors including islet-cell adenomas, neuroendocrine carcinomas of the gastrointestinal and bronchial tracts, and medullary carcinomas of the thyroid (Wiedenmann et al., 1986).

Receptors for tetanus toxin and for the monoclonal antibody A2B5, which identify specific gangliosides present on nerve cells and astrocytes, have also been found in the endocrine pancreas (Eisenbarth et al., 1982).

Similarly Teitelman and Lee (1987) reported that tyrosine hydroxylase (TH) the first enzyme in the catecholamine (CA) biosynthetic pathway, is transiently produced by all pancreatic islet cells but never in acinar exocrine cells.

Moreover, a detailed lineage analysis of the endocrine cells of the pancreas has been performed during ontogeny of transgenic mice carrying hybrid genes in which the 5' regulatory sequence of the rat insulin gene was linked to the coding sequence of simian virus (SV40) large T antigen (Tag) (Alpert et al. 1988).

Alpert et al. found that islet cells synthesizing TH plus glucagon, somatostatin, or pancreatic polypeptide coexpressed the transgene when they first arose. A developmental lineage is proposed for the pancreatic islet cells on the basis of coexpression at an early stage of genes, the activity of which is later restricted to distinct cell types. A precise chronological sequence is established for the differentiation of all defined islet cell types from a common pluripotent ancestor, able to transcribe not only the glucagon and insulin genes but also the TH gene.

Surprisingly, these authors also found that Tag was present transiently in the developing nervous system specifically, in some neural crest cells and in the basal plate of the neural tube, the rhombencephalon, rostral mesencephalon and diencephalon. It is noteworthy that monoaminergic systems, producing serotonin and catecholamines, subsequently develop in some of these locations. These observations reinforce the long-standing suggestion that the islet cells may be derived from neurectodermal precursors rather than from the endoderm.

However, although the list of molecular markers common to the nervous system and the polypeptide-hormone-producing cells of the APUD series is impressive and may be suggestive that these cells share a common ancestor, this cannot be taken as proof since there is no obligatory ancestral relationship between cells expressing the same structural genes.

Definitive evidence for kinship within a common lineage for diverging histiotypes can come only from either clonal analysis, if the developmental potentialities of embryonic cells are evaluated in single cell-derived colonies *in vitro* (e.g. Baroffio et al., 1988; Dupin et al., 1990 for the neural crest, see also *in vivo* studies, Bronner-Fraser and Fraser, 1988, 1989), or from the use of embryological cell tracing techniques.

1.3 Cell Tracing Methods Applied to the Problem of the Origin of the DNES

In the 1970s, using the newly developed quail-chick marker system, we have carried out studies to determine which, if any, of the APUD cells could be derived from precursors in the neural crest. No doubt remained about the ectodermal origin of the adrenal medulla, or of the pituitary gland, since the hypophyseal placode (which later becomes the Rathke's pouch) appeared closely related in its mode of formation to the neural anlage. Recently, its presumptive territory at early somitic stages has been very precisely localized to the most anterior region of the neural fold, in continuity with the neural plate area destined to form the hypothalamus (Couly and Le Douarin, 1985, 1987).

By using the quail-chick chimera system, the neurectodermal origin of carotid body cells and of the glandular components of the ultimobranchial body (UB), the role of which is to produce calcitonin (C-cells), proposed by Pearse,

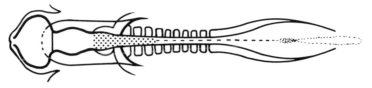

Fig. 1-1 Chimeric embryos were constructed by substituting the rhombencephalon of a chick by its counterpart at stage 6- to 10-somites.

was confirmed. Convergent data from our group and Ann Andrew's laboratory however, invalidated the hypothesis of a possible derivation of the pancreatic islets and of the hormone producing cells of the gut epithelium from the neural crest.

The carotid body which plays a role as chemoreceptor is a paired glandular structure located close to the carotid artery and parathyroid glands. Investi-

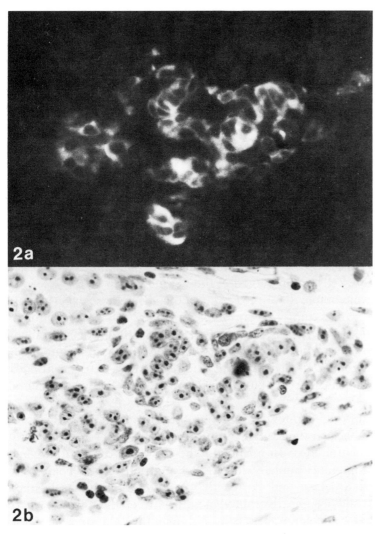

Fig. 1-2 Carotid body of a 14-day-old chick embryo which had received at the 10 somite stage a graft of a quail rhombencephalon. a. FIF technique shows a bright greenish fluorescence in the gland. b. The Feulgen and Rossenbeck staining shows that all the fluorescent cells have the quail nuclear marker.

gations on the origin of the carotid body based mainly on the classical techniques of descriptive embryology had produced until the sixties controversial results. According to Rogers (1965) the first visible sign of the developing carotid body is a "primary condensation of cells" of unknown origin "on the third aortic arch".

Type I cells that contain dense core granules, when subjected to the APUD-FIF procedure, show positive fluorescence essentially due to 5HT in chicken (with a yellowish colour) and dopamine in the quail (green fluorescence) as shown by Pearse et al. (1973).

Chimeric embryos were constructed by substituting the rhombencephalon of a chick by its quail counterpart at stage 6 to 10-somites (Fig. 1-1). The neural crest cells migrating from the graft invaded the branchial and glandular structures (Le Douarin et al., 1974; Le Lièvre and Le Douarin, 1975; Le Lièvre, 1974). Under these experimental conditions the carotid bodies (except for endothelial cells) were entirely derived from the grafted neural crest. When treated consecutively for FIF and for Feulgen nuclear staining procedures, the carotid body cells of such chimaeras exhibited a bright greenish fluorescence characteristic of quail type I cells and the heterochromatin rich nucleoli of the quail nuclei (Le Douarin et al. 1972; Pearse et al. 1973) (Fig. 1-2). Microspectrofluorometric analysis of the biogenic amine content of the carotid body type I cells of the chimaeras revealed the presence of dopamine as in the normal quail carotid body. At the electron microscope level, the type I cells contained small dense core granules and had the quail nuclear marker indicating that these cells originated from the neural crest.

Similar results were obtained for the C cells of the UB in these experiments. The C cells of the grafted embryos were characterized by 1) the observation of formol induced fluorescence resulting from the presence of fluorogenic amines 2) the presence at the electron microscopic level of electron dense core granules and 3) a positive immunocytochemical reaction to anti-calcitonin antibody along with the quail nuclear marker attesting their origin from the neural crest.

In mammals, a different strategy was used to trace the parafollicular cells of the thyroid glands back to their origin from the branchial arch mesectoderm derived from the neural crest. In mammals, the UB joins the thyroid gland when its rudiment migrates caudally during neck morphogenesis (Pearse and Carvalheira, 1967; Stoeckel and Porte, 1970). Ultimobranchial cells then invade the thyroid and become distributed mainly between the follicles, where they are termed parafollicular or clear cells, although they are sometimes also inserted into the follicular epithelium itself.

Cells with APUD characteristics have been described in the early mouse embryo in the vicinity of the fourth pharyngeal pouch and have been claimed to be C-cell precursors on their way to the ultimobranchial body endoderm

(Pearse and Polak, 1971), without proof, however, of their neural crest origin or ulterior fate. One of us (Fontaine, 1979) reinvestigated this question. The pharyngeal region of 18- to 45-somite mouse embryos was dissected, in order to isolate the thyroid rudiment (Tab. 1-II, series A).

In a separate series of animals, the thyroid rudiment remains associated with the last pharyngeal pouches (Tab. 1-II, series B,C,D). In both cases, the pharyngeal regions so defined were transplanted for about 14 days on the chorioallantoic membrane of chick embryos used as a culture "medium". In series B (Tab. 1-II) the endoderm and the mesenchyme of the thyroid and branchial pouches were included in the explants, while in series C the mesenchyme was removed at the level of the branchial pouches; finally in series D the mesenchyme was present but the endoderm of the last pouch was removed.

C-cell differentiation was detected in the explants by application of the FIF technique after L-DOPA injection and also by electron microscopic observation of the tissues. No C-cells developed in the thyroid rudiment explanted

Tab. 1-2 Different types of explants and the presence of C-cells in the thyroid gland after 14 days in graft on the chorioallantoic membrane of 6-day-old chick embryos.

Series	Type of explant	Stages of operation			Results	
		From 17 to 25 somites	From 25 to 28 somites	From 30 to 45 somites	C-cell differentiation +	−
A	BP 1st / 2nd / 3rd / 4th	9*				25 (100%)
			7			
				9		
B		6				23 (100%)
			10			
				7		
C		17			4 (24%)	13 (76%)
				39	39 (100%)	
D		48			41 (86%)	7 (14%)
			11		5 (49%)	6 (51%)
				8		8 (100%)

Endoderm □   Mesenchyme ▨   BP, branchial pouch

* Numerals indicate number of cases throughout

alone (Fig. 1-3), whereas they were present in every case when the complete ultimobranchial body primordium was included in the graft (Fig. 1-4). Series C and D indicated in addition that the precursor C-cells were mainly distributed in the mesenchymal component of the branchial arch until the 28- somite stage. Thereafter, and during a short period of time corresponding to 28- to 30-somite

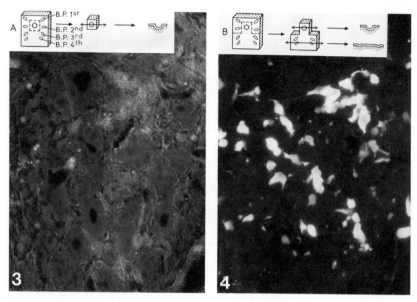

Fig. 1-3 Thyroid tissue resulting from the evolution of the median thyroid bud (endoderm + mesenchyme) of a 25-somite mouse embryo grafted during 14 days on chorioallantoic membrane of a 6-day-old chick embryo. Treatment by the FIF technique after injection of L-DOPA. No fluorescent cells are present in the gland.

Fig. 1-4 Result of the graft of the median thyroid (endoderm + mesenchyme) together with the last pharingeal pouches (endoderm + mesenchyme) from a 25-somite stage mouse embryo. Numerous fluorescent C cells are present in the thyroid. FIF technique after injection of L-DOPA.

Fig. 1-5 Thyroid developed in graft from an explant comprising the median thyroid bud ▶ (endoderm + mesenchyme) and the endoderm of the third and fourth branchial pouches of a 20-somite mouse embryo. The rudiment was grafted for 14 days on the chorioallantoic membrane of a 6-day-old chick embryo. (a) The glandular tissue was totally deprived of cells containing fluorogenic amines. FIF technique after injection of L-DOPA. (b) The same section stained with PAS-hematoxylin showed numerous colloid-containing follicles.

Fig. 1-6 Same experiment as in Fig. 1-5 carried out at 31-somite stage. (a) The thyroid gland contained fluorescent cells. (b) Same section stained with PAS-hematoxylin. A few thyroid follicles have differentiated.

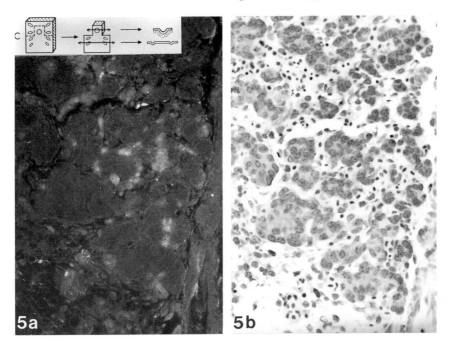

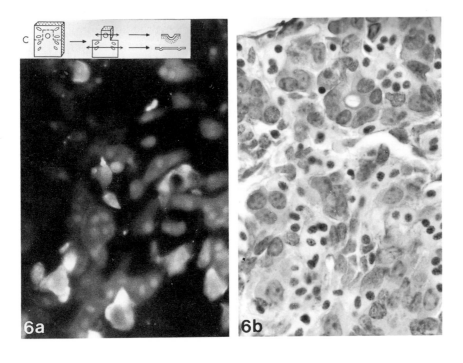

stages, the presumptive C-cells invaded the endoderm of the last pouch where they remained until the UB became confluent with the thyroid (Figs. 1-5, 1-6, 1-7, 1-8). Since the branchial arch mesenchyme has been shown to originate mostly from the neural crest in mammals as well as in birds (Johnston, 1966), one can conclude from these experiments that the final localization of C-cells in the thyroid gland involves a multistep migration, first from the neural primordium to the branchial arch mesenchyme, then from the latter to the endoderm of the ultimobranchial rudiment, and finally from the ultimobranchial body itself to the developing thyroid.

These results support Pearse's hypothesis of the non-endodermal but neural crest origin of the C-cells in mammalian as well as in bird embryos.

Interestingly, Barash et al. (1987) have recently shown that cultured C-cells from adult sheep thyroid respond to the β subunit of nerve growth factor by extending neurites and switching expression from calcitonin to calcitonin-gene-related peptide (CGRP), a peptide also found in a number of neuronal cell types in both the central and peripheral nervous systems. C-cells therefore share with adrenomedullary cells a common origin from the neural crest and a common response to nerve growth factor.

In contrast with the above mentioned cell types (carotid body type I cells and C-cells), the hypothesis of a neural crest origin has been shown to be untenable for the endocrine cells of the pancreas and of the gastrointestinal mucosa that secrete polypeptide hormones or neuropeptides.

From descriptive studies, which can only yield inconclusive data, endocrine cells were claimed to be derived from ganglion cells of the autonomic system (Danisch, 1924; Chung, 1934). But Simard and Van Campenhout (1932) observed that no nervous elements are present in the intestine of chick embryos prior to 32 hours of incubation, whereas intestinal endocrine cells differentiated in chorioallantoic grafts of gut primordia taken from 82 hour embryos. Andrew (1963, 1974) isolated from the head-process- to the 22-somite-stage chick embryos, embryonic territories with gut potencies plus or minus the neural crest area. After culture on the chorioallantoic membrane, most of the explants

---

Fig. 1-7 Thyroid developed in graft from an explant comprising the median thyroid bud ▶ (endoderm + mesenchyme) and the mesenchyme of the third and fourth pharyngeal pouches of a 22-somite mouse embryo (14 days in chorioallantoic membrane graft). (a) The thyroid shows fluorescent C-cells, some of which were inserted in the follicle epithelium. FIF technique after injection of L-DOPA. (b) The same section stained with PAS-hematoxylin reveals that the glandular tissue differentiated normally.

Fig. 1-8 Same experiment as in Fig. 1-7 carried out a 30-somite stage. (a) The glandular tissue does not contain fluorescent cells. FIF technique after injection of L-DOPA. (b) The same section stained with PAS-hematoxylin.

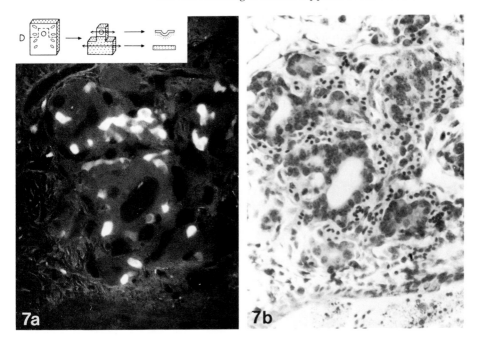

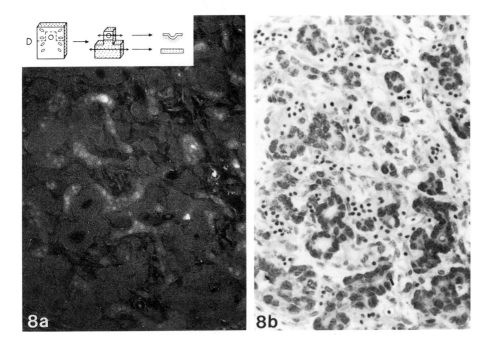

exhibited well-developed intestinal structures, all showing enterochromaffin cells, while enteric ganglia, were present only in explants that included the crest. These experiments strongly suggested that the neural crest is not the source of enterochromaffin cells.

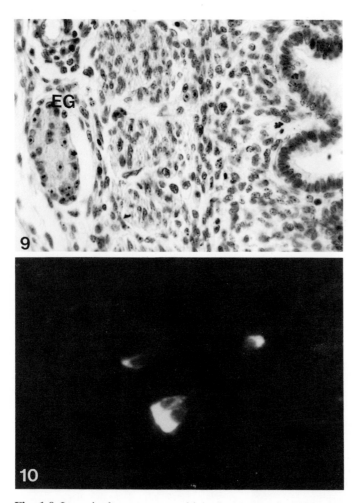

Fig. 1-9 Intestinal structures which developed in the explants composed of chick endomesoderm and quail ectoblast of quail. The association of the two germ layers was performed at stage 6 of Hamburger and Hamilton and the explant was cultured for 14 days. Quail cells originating from the neurectoderm have migrated into the gut structures and have given rise to the enteric ganglia (EG), but no quail cells are seen in the gut epithelium. Feulgen-Rossenbeck staining.

Fig. 1-10 Immunocytochemical localization of somatostatin. Same experiment as in Fig. 1-9. The gut epithelium contains immunoreactive endocrine cells.

By means of isotopic and isochronic transplantations of fragments of the quail neural primordium into chick embryos (or vice versa), at all the levels of the neural axis Le Douarin and Teillet (1973) showed that, although cells do migrate into the gut from the neural crest (two privileged regions, the "vagal" level located between somites 1 to 7, and the "lumbosacral" area caudal to the level of somite 28 give rise to these gut-seeking cells), none reaches the endodermal epithelium. Rather the invading crest cells differentiate into the enteric ganglia of Auerbach's and Meissner's plexuses; they do not colonize the endoderm, which nevertheless contains cells with APUD characteristics (Figs. 1-9, 1-10).

The possibility that these cells might be derived from the neurectoderm at a stage preceeding the onset of the neural crest structure was thereafter tested experimentally (Fontaine and Le Douarin, 1977). The endomesoderm of chick embryos was associated with the ectodermal germ layer of quail blastoderms at various stages including the formation of the primitive streak, the head-process and the neural plate, ranging from 12 to 24 hours of incubation. The recombined embryos were either cultured *in vitro* or on the chorioallantoic membrane and the intestinal structure which developed in the explants was analyzed for chimaerism using various cytochemical techniques: FIF technique after L-DOPA injection, lead haematoxylin, silver staining to indicate argentaffinity and argyrophily, all combined with the Feulgen-Rossenbeck reaction.

In all cases enteric ganglia originated from the quail ectoderm, but the enterochromaffin as well as other APUD cells, which developed normally in the epithelium, were always of chick type. Therefore, no migration of cells from the ectoderm into the endoderm appeared to occur before formation of the neural crest. Thus a neurectodermal origin for the endocrine or paracrine cells of the gastrointestinal tract epithelium must be excluded.

The hypothesis of a neural crest origin for the pancreatic endocrine cells has also been tested in avian embryos subjected to isotopic and isochronic transplantations of the neural primordium. Such grafts were made from quail into chick embryos at the vagal level of the neuraxis. Subsequently small groups of quail cells were found in the pancreatic tissue but did not correspond to the islets of Langerhans (Le Douarin and Teillet, 1973). Fontaine et al. (1977) studied the type of differentiation expressed by these cells and observed that the clusters of quail cells were always separate from both exocrine and endocrine structures. They were subjected to the FIF technique after L-DOPA injection, and their affinity for lead haematoxylin was investigated. The cells originating from the neural crest, identified by the nuclear marker, did not exhibit the cytochemical properties of pancreatic endocrine cells. In fact, the crest cells that had migrated into the pancreas, differentiated into parasympathetic ganglia, as demonstrated by silver impregnation techniques (Fig. 1-11).

Antisera directed against glucagon, insulin and somatostatin, marking respectively A-, B- and D-cells, were applied to the chimaeric pancreas. As in previous assays, the endocrine cells so identified never carried the quail marker (Fontaine-Pérus et al., 1980). The same type of experiment was performed by Andrew (1976), who grafted labelled neural crest cells in chick-quail embryo chimeras and followed their fate in the pancreatic rudiment in 3–4-day hosts. Although pancreatic differentiation is not yet fully expressed at this stage, some cells with APUD characteristics can already be distinguished (Andrew, 1974) in the pancreatic bud, and none of them was ever labelled with the crest cell marker.

Dieterlen-Lièvre and Beaupain (1976) reached similar conclusions through another experimental approach in which fragments of the splanchnopleure

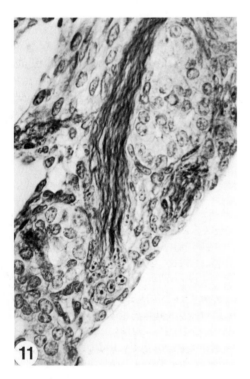

Fig. 1-11 Application of silver impregnation technique on the chimeric pancreas showing that the neural crest cells which migrate into this organ differentiate into nerve cells. Section throught the pancreas of a 14-day-old chick which had received at the 7-somite stage an isotopic graft of quail neural tube at the level of the Ist to the 7th somite. The group of quail cells is associated with a large bundle of nerve fibers.

(endoderm plus mesoderm taken at the presumptive pancreatic level from embryos with from 8 to 15 pairs of somites) were transplanted into the coelomic cavity of 3-day chick embryos. Immunocytochemical localization of insulin- and glucagon-producing cells was performed in the explants, showing that these cell types develop in the absence of neurectodermal cells.

A similar stategy was used in the mammalian embryo by Pictet et al. (1976). Nine-day rat embryos (prior to the 4-somite stage) were treated with a solution of trypsin-pancreatin: thus the whole ectodermal sheet was peeled off. At this time the neural groove has not closed and the neural crest is not yet formed. The mesendoderm was then cultured for 11 days; in every case with a developed pancreas, insulin was detected and B-cells were identified. Therefore, as in other experiments reported above, a neural crest origin of pancreatic islet B-cells is disproven.

1.4 Concluding Remarks

In conclusion, one can say that, in contrast to the unifying hypothesis proposed by Pearse, as tempting as it may be, the polypeptide hormone producing cells distributed in dispersed locations throughout the body do not all share a common embryological origin from the ectodermal germ layer. Certain of these cells are indeed derived from the ridges of the initial neural primordium i.e. from the neural folds either anterior for the pituitary or lateral for those originating from migratory neural crest cells. However, the endocrine cells associated with the digestive tract including the gut epithelium and the pancreas, could not be traced back to the ectoderm as it exists when gastrulation is completed.

Such convergence of differentiation phenotypes arising from different germ layers in fact is not exceptional in embryonic development. One other striking case concerns some bones of the face (e.g. the frontal) that is formed by cells arising both from ectoderm and mesoderm (Le Lièvre, 1971).

1.5 References

1 Alpert, S., Hanahan, D. and Teitelman, G. (1988) Hybrid insulin genes reveal a developmental lineage for pancreatic endocrine cells and imply a relationship with neurons. *Cell*, **53**, 295–308

2 Andrew, A. (1963) A study of the developmental relationship between enteroch- romaffin cells and the neural crest. *J. Embryol. Exp. Morphol.* **11**, 307–324

3 Andrew, A. (1974) Further evidence that enterochromaffin cells are not derived from the neural crest. *J. Embryol. Exp. Morphol.* **31**, 589–598

4 Andrew, A. (1976) An experimental investigation into the possible neural crest origin of pancreatic APUD (islet) cells. *J. Embryol. Exp. Morphol.* **35**, 577–593

5 Barash, J.M., Mackey, H., Tamir, H., Nunez, E.A. and Gershon, M.D. (1987) Induction of a neural phenotype in a serotoninergic endocrine cell derived from the neural crest. *J. Neurosci.,* **7**, 2874–2883

6 Baroffio, A., Dupin, E. and Le Douarin, N.M. (1988) Clone-forming ability and differentiation potential of migratory neural crest cells. *Proc. Natl. Acad. Sci, USA,* **85**, 5325–5329

7 Bayliss, W.M. and Starling, E.H. (1902) On the causation of the so-called "peripheral reflex secretion" of the pancreas. *Proc. Roy. Soc. London,* **69**, 352–353

8 Bernard, C. (1856) Mémoire sur le pancréas. *C. R. Acad. Sci. suppl. I,* Paris, 379–563

9 Bronner-Fraser, S. and Fraser, S. (1988) Cell lineage analysis reveals multipotency of some avian neural crest cells. *Nature,* **335**, 161–164

10 Bronner-Fraser, S. and Fraser, S. (1989) Developmental potential of avian trunk neural crest cells in situ. *Neuron,* **3**, 755–766

11 Chung, I. (1934) Beiträge zur Kenntnis der Entwicklung der gelben Zellen im Darm. *J. Chosen Med. Assoc.* **24** (German summary) 52

12 Cochard, P. and Coltey, P. (1983) Cholinergic traits in the neural crest: acetylcholinesterase in crest cells of the chick embryo. *Dev. Biol.,* **98**, 221–238

13 Copp, D.H., Cameron, E.C., Cheney, E.A., Davidson, A.G. and Henze, K.G. (1962) Evidence for calcitonin. A new hormone from the parathyroid that lowers blood calcium. *Endocrinology,* **70**, 638–649

14 Couly, G.F. and Le Douarin, N.M. (1985) Mapping of the early neural primordium in quail-chick chimeras. I. Developmental relationships between placodes, facial ectoderm and prosencephalon. *Dev. Biol.,* **110**, 422–439

15 Couly, G.F. and Le Douarin, N.M. (1987) Mapping of the early neural primordium in quail-chick chimeras. II. The prosencephalic neural plate and neural folds: implications for the genesis of cephalic human congenital abnormalities. *Dev. Biol.,* **120**, 198–214

16 Danisch, F. (1924) Zur Histogenese der sogenannten Appendix karzinoïde. *Beitr. Pathol. Anat.* **72**, 687–709.

17 Dieterlen-Lièvre, F. and Beaupain, D. (1976) Immunocytological study of endocrine pancreas ontogeny in the chick embryo: normal development and pancreatic potentialities in the early splanchnopleure. In: *The Evolution of Pancreatic Islets* (T.A.I. Grillo, L. Liebson and A. Epple, eds.). Pergamon Press, Oxford, pp. 37–50

18 Drews, U., Kussather, E. and Usadel, K.H. (1967) Histochemischer Nachweis der Cholinesterase in der Frühentwicklung der Hühnerkeimscheibe. *Histochemie,* **8**, 65–89

19 Dupin, E., Baroffio, A., Dulac, C., Cameron-Curry, P. and Le Douarin, N.M. (1990) Schwann cell differentiation in clonal culture of the neural crest as evidenced by the anti-Schwann cell myelin protein monoclonal antibody. *Proc. Natl. Acad. Sci.,* **87**, 1119–1123

20 Eisenbarth, G.S., Shimizu, K., Bouring, M.A. and Wells, S. (1982) Expression of receptors for tetanus toxin and monoclonal antibody A2B5 by pancreatic islet cells. *Proc. Natl. Acad. Sci. USA,* **79**, 5066–5070

21 Falkmer, S., Hakanson, R., and Sundler, F. (1984) In: *Evolution and Tumor Pathology of the Neuroendocrine System* (Falkmer et al., eds.). Elsevier, Amsterdam, New York, pp. 433–452

22 Feyter, F. (1938) *Über diffuse endokrine epitheliale Organe.* J.A. Barth, Leipzig

23 Fontaine, J. (1979) Multistep migration of calcitonin cell precursors during ontogeny of the mouse pharynx. *Gen. Comp. Endocrinol.* **37**, 81–92

24 Fontaine, J. and Le Douarin, N. (1977) Analysis of endoderm formation in the avian blastoderm by the use of quail-chick chimaeras. The problem of the neurectodermal origin of the cells of the APUD series. *J. Embryol. Exp. Morph.,* **41**, 209–222

25 Fontaine, J., Le Lièvre, C. and Le Douarin, N. M. (1977) What is the developmental fate of the neural crest cells which migrate into the pancreas in the avian embryo? *Gen. Comp. Endocrinol.* **33**, 394–404

26 Fontaine-Pérus, J., Le Lièvre, C. and Dubois, P.M. (1980) Do neural crest cells in the pancreas differentiate into somatostatin-containing cells? *Cell Tissue Res.* **213**, 293–299

27 Gregory, R.A. and Tracy, H.J. (1964) The constitution and properties of two gastrins extracted from hog antral mucosa. *Gut,* **5**, 103–117

28 Johnston, M.C. (1966) A radioautographic study of the migration and fate of cranial neural crest cells in the chick embryo. *Anat. Rec.* **156**, 143–156

29 Landis, S.C. and Patterson, P.H. (1981) Neural crest cell lineages. *Trends in Neurosci.,* **4**, 172–175

30 Le Douarin, N. (1982) *The Neural Crest,* Cambridge University Press, Cambridge, 259 pages

31 Le Douarin, N.M., Fontaine, J. and Le Lièvre, C. (1974) New studies on the neural crest origin of the avian ultimobranchial glandular cells. Interspecific combinations and cytochemical characterization of C cells based on the uptake of biogenic amine precursors. *Histochemistry* **38**, 297–305

32 Le Douarin, N.M., Le Lièvre, C. and Fontaine, J. (1972) Recherches expérimentales sur l'origine embryologique du corps carotidien chez les Oiseaux. *C. R. Acad. Sci. Paris,* **275**, 583–586

33 Le Douarin, N.M. and Teillet, M.-A. (1973) The migration of neural crest cells to the wall of the digestive tract in avian embryo. *J. Embryol. Exp. Morphol.* **30**, 31–48

34 Le Lièvre, C. (1971) Recherches sur l'origine embryologique des arcs viscéraux chez l'embryon d'Oiseau par la méthode des greffes interspécifiques entre Caille et Poulet. *C.R. Soc. Biol.,* **165**, 395–400

35 Le Lièvre, C. (1974) Rôle des cellules mésectodermiques issues des crêtes neurales céphaliques dans la formation des arcs branchiaux et du squelette viscéral. *J. Embryol. Exp. Morph.,* **31**, 453–477

36 Le Lièvre, C. and Le Douarin, N.M. (1975) Mesenchymal derivatives of the neural crest: analysis of chimaeric quail and chick embryos. *J. Embryol. Exp. Morph.,* **34**, 125–154

37 Ljungberg, O., Cederquist, E. and Studnitz, W. (1967) Medullary thyroid carcinoma and phaeochrocytoma: a familial chromaffinomatosis. *Br. Med. J.,* 279–281

38 Marangos, P.J., Zis, A.P., Clark, R.L. and Goodwin, F.K. (1978) Neuronal, non-neuronal and hybrid forms of enolase in brain: structural, immunological and functional comparisons. *Brain Res.,* **150**, 117–133

39 Marangos, P.J., Campbell, J.C., Schmechel, D.E., Murphy, D.L. and Goodwin, F.K. (1980) Blood platelets contain a neuron-specific enolase subunit. *J. Neurochem.,* **34**, 1254–1258

40 Mering, J. and Minkowski, O. (1889) Diabetes mellitus nach Pankreas-exstirpation. *Arch. f. Exp. Pathol. u. Pharmakol.*, **26**, 372–387

41 Mutt, V., Jorpes, J.E. and Magnusson, S. (1970) Structure of porcine secretin. The aminoacid sequence. *Europ. J. Biochem.*, **15**, 513–519

42 Pearse, A.G.E. (1968) Common cytochemical and ultrustructural characteristics of cells producing polypeptide hormones (the APUD series) and their relevance to thyroid and ultimobranchial C cells and calcitonin. *Proc. Roy. Soc. London, Ser. B.*, **170**, 71–80

43 Pearse, A.G.E. (1969) The cytochemical and ultrastructure of polypeptide hormone-producing cells of the APUD series and the embryologic, physiologic and pathologic implications of the concept. *J. Histochem. Cytochem.*, **17**, 303–313

44 Pearse, A.G.E. (1976) Peptides in brain and intestine. *Nature*, **262**, 92–94

45 Pearse, A.G.E. (1977) The diffuse endocrine system and the "common peptides". In: *Molecular endocrinology* (I. Mac Intyre and M. Szelke eds.) North-Holland, Elsevier, pp. 309–323

46 Pearse, A.G.E. (1980) APUD: concept, tumours, molecular markers and amyloid. *Mikroskopie*, **36**, 257–269

47 Pearse, A.G.E. and Carvalheira, A. (1967) Cytochemical evidence for an ultimo-branchial origin of rodent thyroid C cells. *Nature*, **214**, 929–930

48 Pearse, A. G. E. and Polak, J. M. (1971) Cytochemical evidence for the neural crest origin of mammalian ultimobranchial C cells. *Histochemie* **27**, 96–102

49 Pearse, A. G. E. , Polak, J. M., Rost, F. W. D., Fontaine, J., Le Lièvre, C. and Le Douarin, N. (1973) Demonstration of the neural crest origin of type I (APUD) cells in the avian carotid body, using a cytochemical marker system. *Histochemistry* **34**, 191–203

50 Pictet, R. L., Rall, L. B., Phelps, P. and Rutter, W. J. (1976) The neural crest and the origin of the insulin-producing and other gastrointestinal hormone-producing cells. *Science* **191**, 191–192

51 Rogers, D. C. (1965) The development of the rat carotid body. *J. Anat.* **99**, 89–101

52 Schmechel, D., Marangos, P.J., and Brightman, M. (1978) Neurone-specific enolase is a molecular marker for peripheral and central neuroendocrine cells. *Nature*, **276**, 834–836

53 Simard, L.C. and Van Campenhout, E. (1932) The embryonic development of argentaffin cells in the chick intestine. *Anat. Rec.*, **53**, 141–159

54 Stoeckel, M. E. and Porte, A. (1970) Origine embryonnaire et différentiation sécrétoire des cellules à calcitonine (cellules C) dans la thyroïde foetale du rat. Etude au microscope électronique. *Z. Zellforsch.* **106**, 251–268

55 Teitelman, G., Joh, T.H. and Reis, D.J. (1981) Linkage of the brain-skin-gut axis: islet cells originate from dopaminergic precursors. *Peptides*, **2**, 157–168

56 Teitelman, G. and Lee, J.K. (1987) Cell lineage analysis of pancreatic islet cell development: glucagon and insulin cells arise from catecholaminergic precursors present in the pancreatic duct. *Dev. Biol.*, **121**, 454–466

57 Wiedenmann, B., Franke, W.W., Kuhn, C., Moll, R. and Gould, V.E. (1986) Synaptophysin: a marker protein for neuroendocrine cells and neoplasms. *Proc. Natl. Acad. Sci. USA*, **83**, 3500–3504

58 Zollinger, R.M. and Ellison, E.H. (1955) Primary peptic ulceration of the jejunum associated with islet cell tumors of the pancreas. *Ann. Surg.*, **142**, 709–728

Part II

Molecular and Cell Biology

2 Membrane Proteins of Synaptic Vesicles:
Markers for Neurons and Endocrine Cells; Tools for the Study of Neurosecretion

Reinhard Jahn and *Pietro De Camilli*

2.1 Introduction

Intercellular signalling in the nervous system is mediated largely by hydrophilic molecules, the neurotransmitters, which are released from presynaptic nerve endings upon stimulation. Two classes of neurotransmitters can be distinguished: small, non-peptide molecules (often referred to as classical neurotransmitters, e.g. acetylcholine, catecholamines and certain amino acids), and the peptides. It was formerly believed that non-peptide neurotransmitters were the only transmitters in most synaptic terminals of the central and peripheral nervous system and that peptides were confined to a small number of specialized neurons. Recently, evidence has accumulated showing that most, and perhaps all, nerve terminals are capable of secreting a variable cocktail of molecules including both classical and peptide neurotransmitters (Hökfelt et. al., 1986).

It is well established that neurotransmitters are stored in secretory vesicles which undergo calcium-dependent exocytosis upon stimulation. At least two types of morphologically distinct secretory vesicles exist in nerve terminals. One type is represented by small, highly homogeneous vesicles, with an average diameter of 50 nm in the mammalian nervous system (Peters et al., 1976). They are referred to as SSVs (small synaptic vesicles) and are thought to be involved only in the storage and secretion of classical neurotransmitters. They are translucent when observed in the electron microscope. However, catecholamine-containing SSVs accquire a dense core under certain fixation conditions

(Klein et al., 1982; Hökfelt et al., 1986). The second type is represented by larger vesicles (70–200 nm diameter) which contain an electron-dense core. These vesicles, referred to as LDCVs (large dense core vesicles), are the storage organelles of neuropeptides but may in addition also contain classical neurotransmitters. LDCVs are particularly concentrated in nerve terminals of certain parts of the brain (e.g. the hypothalamus) but are present in variable

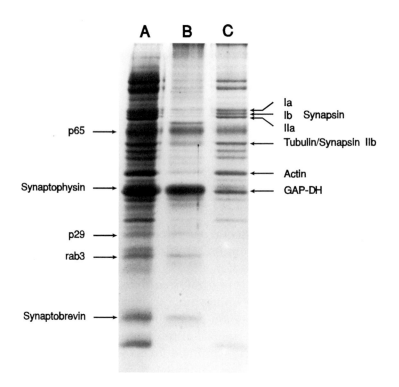

Fig. 2-1 Proteins from mammalian SSVs, separated by SDS-polyacrylamide gel electrophoresis (9–13% gradient gel) and visualized by staining with Coomassie Blue. Lane A shows the profile of highly purified SSVs (method of Nagy et al. (1976) with the modifications by Huttner et al., 1983; see also Fig. 2-2). This was followed by the separation of detergent-binding proteins (lane B, mostly integral membrane proteins) from membrane-associated proteins (lane C) using the Triton X-114 procedure. The proteins were identified with the aid of monospecific antibodies or by N-terminal sequencing (only glutaraldehyde-3-phosphate dehydrogenase [GAP-DH]). The bands corresponding to the positions of synapsin IIa and tubulin/synapsin IIb contain additional unidentified proteins. Note that the G-protein, rab3, codistributes with the integral membrane proteins despite the lack of a membrane-spanning domain which is probably due to a hydrophobic posttranslational modification. See text for details. For methods see Baumert et al. (1989).

number at most synapses of the central and peripheral nervous system and may be present in every nerve cell (for review see Peters et al., 1976; Klein et al., 1982; Hökfelt et al., 1986; Thureson-Klein and Klein, 1990).

In the last few years, significant progress has been made in elucidating the molecular composition of SSVs. Presently, the membrane composition of SSVs is better characterized than that of any other secretory organelle. This is due to the application of modern techniques of immunochemistry and molecular biology. SSVs share a group of abundant membrane proteins with unique structural properties which are highly specific for these organelles (Fig. 2-1). Although the function of none of these proteins is conclusively established, their identification and characterization allow new experimental approaches for the solution of problems not accessible before. For example, the pathways for vesicle biogenesis, recycling and degradation are being studied using these proteins as markers. Some of these studies have led to new concepts concerning the membrane dynamics involved in these processes. In contrast, the data concerning the membranes of LDCVs are still limited and mainly restricted to a few accessible, highly specialized, model systems.

In this review, we will summarize the current information about structurally identified proteins of synaptic vesicles focussing on those present in mammalian SSVs.

2.2 Membrane Proteins of Small Synaptic Vesicles (SSVs)

2.2.1 Preliminary Remarks

An essential prerequisite for the study of synaptic vesicle proteins is the availability of synaptic vesicle fractions in sufficient quantities for biochemical analysis. Purification protocols were first developed for mammalian brain (De Robertis et al., 1963; Whittaker et al., 1963, 1964). Later, synaptic vesicles from the electric organ of electric rays, which are purely cholinergic, were established as a model system and studied in detail (for review see Kelly et al., 1979; Whittaker, 1984; Zimmermann, 1988; Whittaker, 1988). Major efforts were made to evaluate the degree of contamination of these vesicle preparations by other membranes, using acetylcholine content as a marker for SSVs. Acetylcholine storage appears to survive prolonged isolation procedures (Whittaker et al., 1964; Zimmermann, 1982; Whittaker, 1988), in contrast to that of amino acid transmitters (Burger et al., 1989). These studies resulted in essentially pure and homogeneous vesicle preparations (for review see

Fig. 2-2 Electron micrograph illustrating the morphological homogeneity of mammalian SSVs purified as in Fig. 2-1. Although many other purification protocols are available, this preparation is thoroughly characterized and remains a reference standard for the purity of mammalian brain SSV preparations. Reproduced from Huttner et al. (1983) by copyright permission of the Rockefeller University Press.

Zimmermann, 1982). Preparations of similar purity are also available from mammalian brain (e.g. Nagy et al., 1976; Huttner et al., 1983; Hell et al., 1988; Burger et al., 1989; see Figs. 2-2 and 2-3) although these vesicles are probably heterogeneous with respect to their neurotransmitter content.

Two different approaches have been adopted to study synaptic vesicle proteins. First, functional properties of isolated vesicles were investigated using enzyme or transport assays. These studies led to the identification of various activities which are associated mostly with the uptake and storage of

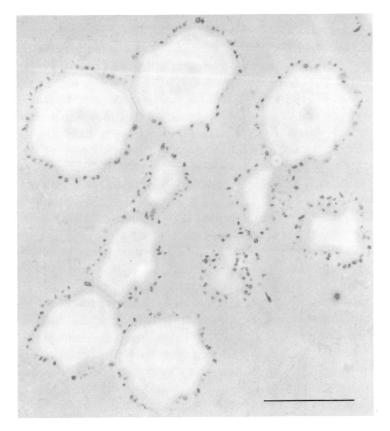

Fig. 2-3 Electron micrograph showing the morphology of SSVs which were isolated by immunoadsorption. Methacrylate microbeads were used as immunosorbent which contained covalently-bound monoclonal antibodies directed against synaptophysin on their surface. This procedure allows the rapid purification of small amounts of SSVs under mild conditions in a fraction of the time required by conventional protocols (Burger et al., 1989). The purity is comparable to that of the preparation of Huttner et al. (1983). Reproduced from Burger et al. (1989) by copyright permission of Cell Press.

neurotransmitters. It was shown that isolated synaptic vesicles contain uptake systems for acetylcholine (Koenigsberger and Parsons, 1980; Giompres and Luqmani, 1980; Diebler and Morot-Gaudry, 1981), monoamines (for review see Philippu and Matthaei, 1988), glutamate (Disbrow et al., 1982; Naito and Ueda, 1983), GABA (Hell et al., 1988; Fykse and Fonnum, 1988) and glycine (Kish et al., 1989) which are energized by an electrochemical proton gradient (for review see Njus et al., 1986; Marshall and Parsons, 1987; Philippu and Matthaei, 1988; Johnson, 1988; Maycox et al., 1990). This gradient is generated by a proton pump of the vacuolar type present in the vesicle membrane (see section 2.2.9). In addition, catecholamine containing SSVs appear to contain proteins involved in their biosynthesis, notably cytochrome b_{561} and dopamine β-hydroxylase (see section 2.3.2). Cholinergic vesicles were studied in more detail and possess, in addition to the proton pump and the acetylcholine carrier, a transport system for ATP which is atractyloside sensitive (Luqmani, 1981). Furthermore, an ATP-dependent Ca^{++} uptake system (Israel et al., 1980; Michaelson et al., 1980), ion channels (Rahamimoff et al., 1988) and various ATPase activities have been reported in studies on SSVs. It is unclear, whether these activities represent vesicular enzymes or are due to contamination by other subcellular fractions (see e.g. Whittaker, 1988) or, in the case of the ATPases, represent atypical states of the vesicular proton pump (see section 2.2.9). Most of this work was covered exhaustively in several recent reviews to which the reader is referred (Kelly et al., 1979; Klein et al., 1982; Whittaker, 1982; Reichardt and Kelly, 1983; Zimmermann, 1988; Whittaker, 1988). With exception of the proton pump (section 2.2.9), cytochrome b_{561} (section 2.3.2), dopamine β-hydroxylase (section 2.3.2) and candidates for the monoamine carrier (section 2.3.2) and the ATP-transporter (section 2.2.10.2), none of the proteins responsible for these transport activities has been identified.

Second, major proteins of SSV membranes were characterized in order to identify molecules of functional significance. It is clear that all specific interactions of SSVs with other cellular components during exo-endocytotic membrane cycling must be mediated by specific proteins present on their surface or in their membrane. Such proteins may therefore possess specific features responsible for binding of calcium, interaction with the synaptic cytoskeleton, recognition of the plasma membrane, membrane fusion etc.. However, none of the proteins responsible for such activities have been identified with certainty. For these reasons, structural characterization of the major SSV membrane proteins should allow the development of improved tools for functional studies.

The structural characterization of SSV proteins was largely based on the biochemical analysis of purified SSV fractions. In addition, some SSV proteins were discovered by serendipidy in primarily unrelated studies (e.g. synapsin I or p65). In a variety of studies, synaptic vesicle proteins were analyzed by one and two-dimensional electrophoresis or lectin binding to identify their main constituents (e.g. Stadler and Tashiro, 1979; Zisapel and Zurgil, 1979; Zanetta et al., 1981; Smith and Loh, 1981; March et al., 1987). Although these studies provided useful information about some major proteins, they are difficult to compare with each other. In addition, the conclusions in some of these studies were compromised by inherent problems of vesicle purity, by artificial adsorption of proteins during SSV isolation (e.g. proteins of the cytoskeleton: Zisapel et al., 1980; see Fig. 2-1) and by difficulties in identifying individual components in impure fractions. We will therefore restrict our discussion to identified proteins which meet at least some of the following criteria:

– monospecific antibodies or other specific probes are available for their study
– their presence on SSVs is demonstrated by immunocytochemistry at the electron microscopical level
– their interaction with the vesicle membrane is sufficiently specific to rule out artificial association
– they copurify with other synaptic vesicle markers using state-of-the-art vesicle purification protocols
– structural data are available from peptide or cDNA sequencing.

2.2.2 Synapsins

The synapsins are represented by a family of four homologous proteins, synapsin Ia and Ib and synapsin IIa and IIb (previously referred to as protein IIIa and IIIb) which are associated with the cytoplasmic surface of SSVs. They were first identified as major substrates for endogenous phosphorylation activity in the mammalian brain (Johnson et al., 1972; Ueda et al., 1973; Forn and Greengard, 1978; Huang et al., 1982). Subsequent studies led to the concept that they act as a link between SSVs and an actin-based cytomatrix of nerve terminals, thus controlling the availability of SSVs for exocytosis (for review of the older literature see Nestler and Greengard, 1984; De Camilli and Greengard, 1986).

2.2.2.1 Structural Properties

Synapsin Ia and Ib (collectively referred to as synapsin I) are major proteins of SSVs, representing approximately 6% of vesicle protein and 0.4 % of total protein of rat cerebral cortex (Goelz et al., 1981; Huttner et al., 1983). They are two highly homologous, extremely basic (pI > 10), acid soluble and elongated proteins with apparent molecular weights of 86,000 and 80,000 daltons, respectively (Ueda and Greengard, 1977). They consist of a head region and a collagenase-sensitive tail region responsible for the basic pI of the molecule (Ueda and Greengard, 1977). Synapsin I is phosphorylated by three different protein kinases at multiple sites. The head region contains a serine residue that is phosphorylated both by cAMP-dependent protein kinase and by Ca^{++}-calmodulin dependent protein kinase type I (Huttner et al., 1981; Kennedy and Greengard, 1981; Nairn and Greengard, 1987). The sequence preceeding this site (Arg-Arg-Arg-Leu-Ser(P)) conforms well with that of the consensus-sequence for cAMP-dependent protein kinase (Czernik et al., 1987). The tail of the molecule contains two additional phosphorylation sites which are phos-phorylated exclusively by Ca^{++}-calmodulin-dependent protein kinase type II (Huttner et al., 1981; Kennedy and Greengard, 1981; Kennedy et al., 1983). The amino acids around these sites have been determined (Ala-Thr-Arg-Gln-Ala-Ser(P), and Pro-Ile-Arg-Gln-Ala-Ser(P)) and correspond to the minimal requirements for this kinase (Pearson et al., 1985; Czernik et al., 1987).

Synapsin IIa and IIb (previously protein IIIa and IIIb; collectively referred to as synapsin II) are two closely related polypeptides with apparent molecular weights of 74,000 and 55,000 daltons, respectively (Huang et al., 1982; Browning et al., 1987). Like synapsin I, the synapsin II molecules are elongated and acid-soluble, but their isoelectric point is neutral (around 7) and they lack the collagenase-sensitive tail. Synapsin II is phosphorylated by cAMP depen-dent protein kinase and Ca^{++}-calmodulin-dependent protein kinase type I (Huang et al., 1982) at a serine residue corresponding to the phosphorylated serine in the head region of synapsin I. Synapsin II shares with synapsin I the sequence surrounding this serine (Südhof et al., 1989a, see below). In contrast, both synapsin IIa and IIb are not phosphorylated by Ca^{++}-calmodulin-dependent protein kinase type II (Browning et al., 1987; Südhof et al., 1989a).

Recently, the primary structure of the four synapsins has been elucidated (Südhof et al., 1989a). Synapsin Ia and Ib are encoded by two mRNAs that are derived from the same primary transcript by alternative splicing. In accordance with this, synapsin Ia and Ib are encoded by a single copy gene (Südhof et al., 1989a) which is probably located on the X-chromosome (Yang-Feng et al., 1986). The mRNAs encode for two proteins of 704 (rat synapsin Ia) and 668 (rat Ib) amino acids which are identical in their first 659 amino acid residues. The C-terminus is divergent which is due to a 38 nucleotide insertion present in the synapsin Ia mRNA resulting in a frameshift of the remaining common nucleotide sequence. Both proteins are highly conserved within mammalian species, with 96% identical amino acid residues between rat and cow (Südhof et al., 1989a).

Synapsin IIa and IIb are also identical in their N-terminal part and diverge in their C-terminus. Again, the two forms are generated by alternative splicing of a primary transcript derived from a single copy gene although the splicing mechanism appears to be different from that of synapsin I mRNA. In accordance with the biochemical properties of the isolated proteins, synapsin IIa and IIb are smaller in size than synapsin Ia and b, with 586 and 479 amino acid residues, respectively (Südhof et al., 1989a).

A comparison of the primary structure of the four synapsins revealed that the proteins are composed of a mosaic of common and individual domains (Südhof et al., 1989a; Fig. 2-4). Synapsin I and II contain a large homologous region at the N-terminal part which can be divided into three domains. The amino terminal domain (A) contains the phosphorylation site common to all proteins and is highly homologous. This is followed by a stretch (B) of weaker homology which is rich in small amino acids such as Ala and Ser. The third domain (C) is the largest, displaying again high homology among all four proteins. This domain appears to contain the regions responsible for interaction with actin and synaptic vesicles (see below). It is both hydrophobic and highly charged and may form multiple amphipatic α-helices and ß-sheets. There are several stretches of hydrophobic amino acids flanked by charged residues which may be responsible for the insertion of synapsin I into the hydrophobic phase of phospholipid bilayers (Benfenati et al., 1989a; Benfenati et al., 1989b, see below).

The C-terminal parts of the synapsins are variable (Fig. 2-4). Synapsins Ia and Ib share a long, proline-rich, collagenase-sensitive sequence (D) which

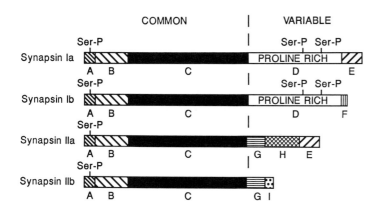

Fig. 2-4 Domain model for the synapsins. See text for details. Reproduced from Südhof et al. (1989a) by copyright permission of the AAAS.

contains the two phosphorylation sites for Ca^{++}-calmodulin-dependent protein kinase type II. This is followed by the domains E and F which are different due to alternative splicing (see above). Synapsins IIa and IIb have only a short common sequence (G) which is rich in proline, followed by their alternatively spliced carboxy-termini (H,E,I). Interestingly, the carboxy-terminal domain of synapsin IIa (E) is highly homologous to the carboxy-terminal domain of synapsin Ia (E) with the last 18 amino acids being identical.

A comparison of the synapsin sequences with that of cytoskeletal proteins, in particular villin, profilin, gelsolin, protein 4.1 or other actin-binding proteins revealed no significant homologies. This is in contrast to previous suggestions by other laboratories (Baines and Bennet, 1985; McCaffery and De Gennaro, 1986) and confirms that the synapsins represent a separate protein family.

2.2.2.2 Interaction of Synapsin I with SSVs and the Cytoskeleton

The interaction of synapsin I with the membrane of SSVs and phospholipid vesicles has been characterized in detail (Schiebler et al., 1986; Benfenati et al., 1989a; Benfenati et al., 1989b). Synapsin I binds reversibly and with high affinity (K_d approx. 10 nM) to purified synaptic vesicles (Schiebler et al., 1986; Steiner et al., 1987). Phosphorylation on the tail weakens this interaction (Huttner et al., 1983; Schiebler et al., 1986). The binding to SSVs appears to involve two different components. First, it was shown that the head of the molecule interacts strongly with pure phospholipid bilayers when acidic phospholipids are present (Benfenati et al., 1989a). Parts of the molecule (probably located in domain C) penetrate into the hydrophobic core of the membrane. While this interaction does not require an acceptor protein, a second binding component was identified which is dependent on a still unknown protein component in the SSV membrane (Benfenati et al., 1989b). This interaction involves the tail of the molecule and is phosphorylation-dependent. However, it remains to be established which domain of the synapsin I molecule is primarily responsible for the specificity of the interaction with SSV membranes. The protein-dependent binding component may be involved suggesting that the tail conveys specificity. However, this does not explain the specificty of the association of synapsin II, which lacks the tail domain D, with SSV membranes (Browning et al., 1987).

In addition to membrane binding, the interactions of synapsin I with various cytoskeletal proteins, particularly actin, have been characterized. Dephospho-synapsin I bundles actin filaments (Bähler and Greengard, 1987; Petrucci and Morrow, 1987). This effect is reduced upon phosphorylation of the head of the molecule and is virtually abolished when synapsin I is phosphorylated at the tail sites or at all three sites (Bähler and Greengard, 1987). A detailed analysis of actin binding and actin bundling using different fragments of the synapsin I molecule revealed that the head contains at least two different actin binding sites and that the tail is required for bundling. The precise mechanism of actin bundling remains to be established. It does not appear to involve self-association of synapsin I (Bähler et al., 1989a).

The interaction of synapsin I with other cytoskeletal proteins is less well defined. Binding has been reported to occur with spectrin (Baines and Bennet, 1985), neurofilaments (Goldenring et al., 1986) and microtubules (Baines and Bennett, 1986; Petrucci and Morrow, 1987), the latter apparently under phosphorylation control (Petrucci and Morrow, 1987). The interaction of synapsin II with cytoskeletal proteins has not been investigated so far. The similarities between synapsin I and synapsin II in domain C suggest that synapsin II, like synapsin I, might interact with actin.

2.2.2.3. Proposed Functions

Although the functions of the synapsins are still not fully understood, a concept has evolved which assigns a central role to these proteins in the synaptic life cycle of SSVs (De Camilli and Greengard, 1986). According to this model, synapsin I (and, by analogy, also synapsin II) is involved in the regulation of the traffic of SSVs in the nerve terminal by forming a reversible link between synaptic vesicles and an actin-based cytomatrix. Recently, crosslinking structural elements have been visualized directly using cryo electron microscopy (Landis, 1988; Landis et al., 1988; Hirokawa et al., 1989). They are thought to be responsible for the clustering of synaptic vesicles in proximity to the release sites. An increase in the phosphorylation state of synapsin I leads to a dissociation of the vesicles from the actin network (see above; Sihra et al., 1989), allowing them to move to the plasma membrane and undergo exocytosis. Thus, phosphorylation of synapsin I would regulate the amount of vesicles available for exocytosis in response to an incoming action potential. This model is supported by a variety of different studies which show that all physiological or pharmacological manipulations that trigger or facilitate neurotransmitter release cause a rapid and transient phosphorylation of the synapsins (for review see Nestler and Greengard, 1984; De Camilli and Greengard, 1986). Synapsin I phosphorylation does not appear to be directly involved in the exocytotic process itself but rather in making more vesicles available for fusion when increased neurotransmitter release is required. This concept was supported considerably by microinjection studies using the squid giant synapse as a model system (Llinas et al., 1985). Injection of dephospho-synapsin I led to an inhibition of transmitter release without affecting presynaptic currents. In contrast, the phosphorylated forms were without effect. Injection of Ca^{++}-calmodulin-dependent protein kinase type II (i.e. the kinase which phosphorylates the tail region of synapsin I) did not trigger neurotransmitter release but led to a significant increase of transmitter release in response to a depolarizing stimulus.

2.2.3 Synaptophysin

Synaptophysin (also referred to as p38) is the major integral membrane protein of SSVs (Jahn et al., 1985; Wiedenmann and Franke, 1985, see also Fig. 2-1). It was discovered independently by several laboratories using immunochemical methods (Bock and Helle, 1977 [named synaptin in this study: Gaardsvoll et al., 1988]; Jahn et al., 1985; Wiedenmann and Franke, 1985; Obata et al., 1986; Devoto and Barnstable, 1987; Leclerc et al., 1989) due to its high immunogenicity. Numerous monoclonal and polyclonal antibodies have been generated, and some of them are commercially available. It is an abundant protein within the brain, corresponding to 0.3% of total cerebral cortex protein or 7% of synaptic vesicle protein (Knaus et al., 1986; Jahn et al., 1987). In addition, synaptophysin is expressed by a variety of peptide-secreting endocrine cells and related tumors (see section 2.4.2). At the present time, synaptophysin is one of the best studied vesicle proteins. It is widely accepted as standard cytochemical marker for nerve terminals and for neuroendocrine cells as well as for tumors derived from corresponding tissues, but its function is still not understood.

2.2.3.1 Structure and Orientation in the Vesicle Membrane

Synaptophysin is an acidic membrane protein (pI approximately 4.8) with an apparent molecular weight (monomer) of 38,000 daltons (Jahn et al., 1985; Wiedenmann and Franke, 1985). It is N-glycosylated (Rehm et al., 1986), with a slight difference between the neuronal and endocrine form in the sugar part of the molecule (Navone et al., 1986). The primary structure of synaptophysin has been determined (Südhof et al., 1987; Leube et al., 1987; Buckley et al., 1987). The mRNA encodes for a protein of 307 amino acids. Based on hydrophilicity analysis of the amino acid side chains, a structural model was proposed (Südhof et al., 1987; Leube et al., 1987; see Fig. 2-5) which was recently confirmed by fragment analysis using domain-specific antibodies (Johnston et al., 1989a). Synaptophysin has four transmembrane domains which are separated by short

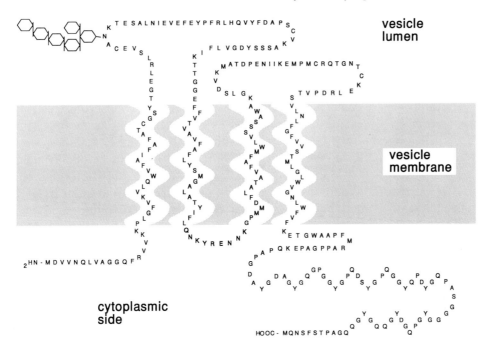

Fig. 2-5 Model of the transmembrane structure of bovine synaptophysin. Amino acids are given in single letter code. Modified from Südhof et al. (1987).

hydrophilic sequences. The transmembrane domains contain stretches of limited sequence homology, suggesting that they were formed by gene duplication events. Both the C and the N termini are facing the cytoplasmic side of the vesicle. The N-terminal part on the cytoplasmic side is short, containing approx. 18 amino acid residues. No signal sequence is observed. The intravesicular loops contain two potential N-glycosylation sites, but apparently only the one located on the first loop is used (Johnston, Jahn and Südhof, unpublished observations). Similarly, cysteines are found only in the domains inside the vesicle. The cytoplasmic C-terminal domain is about 90 amino acids long. It carries most of the antigenic sites (Jahn et al., 1985; Johnston et al., 1989a) and contains 10 copies of an unusual pentapeptide repeat (consensus sequence: Tyr-Gly-Pro(Gln)-Gln-Gly). No homologies to other proteins were found when data banks were searched. However, the tail has some superficial similaritiy with keratin and synexin (Creutz et al., 1988), but more significantly, with octopus rhodopsin. The latter possesses a very similar pentapeptide repeat in its C-terminal domain (8 repeats, consensus sequence: Tyr-Pro-Pro-Gln-Gly; Ovchinnikov et al., 1988).

Synaptophysin has been sequenced from rat, human, cow and *Torpedo* electric organ (Südhof et al., 1987; Leube et al., 1987; Johnston et al., 1989a;

Cowan et al., 1990), allowing a comparison of conserved and variable domains. In the mammalian species, synaptophysin is highly conserved, with 88% identity between all three species (Johnston et al., 1989a). The changes are concentrated in the two intravesicular domains. In contrast, *Torpedo* synaptophysin is more divergent, with only 62% amino acid similarity with rat synaptophysin (Cowan et al., 1990). Interestingly, the tail is least conserved but apparently preserves a gross similarity in its overall structure (rich in Tyr and Pro). The longest stretches of identical amino acids (14 and 12 residues) were found at the aminoterminal region of the fourth transmembrane domain and in the first intravesicular loop, respectively. It will be of interest to determine the primary structure in more distant species (e.g. invertebrates) in order to learn more about the functional constraints of the sequence in evolution.

The subunit structure of the protein is still unclear. Depending on the experimental conditions, synaptophysin behaves as a dimer (Wiedenmann and Franke, 1985; Jahn et al., 1987), trimer (Rehm et al., 1986) or hexamer (Thomas et al., 1988a, see below) formed from identical subunits. Since the cysteine residues are rapidly oxidised, resulting in artificial oligomerization, more work is required to determine the precise structure and to evaluate whether disulfide bridges are involved.

2.2.3.2 Functional Properties

As already mentioned, the biological role of synaptophysin is unknown. However, the protein exhibits several unique properties which may be of functional importance.

Synaptophysin is phosphorylated on tyrosine residues by endogenous tyrosine kinase activity (Pang et al., 1988a). In fact, the protein appears to be one of the few major endogenous substrates for tyrosine kinases in brain (Hirano et al., 1988). This property is shared with other vesicle proteins (Pang et al., 1988a), notably p29 (see below; Baumert et al., 1990). The kinase responsible for this phosphorylation appears to be c-src which is enriched in vesicle fractions (Pang et al., 1988b; Barnekow et al., 1990 in press).

In addition, it was reported that the cytoplasmic tail binds Ca^{++}-ions (Rehm et al., 1986), making it a possible candidate transducing the Ca^{++} signal into exocytosis of SSVs. Furthermore, it was demonstrated that synaptophysin forms an ion channel when incorporated into planar lipid membranes (Thomas et al., 1988a). The channel displayed an average conductance of 154 pS, with the open probability being linearly dependent on the membrane potential. Addition of a tail-specific monoclonal antibody led to alterations of the channel properties. The functional significance remains to be established. Alternative functions have been proposed for synaptophysin such as an organizing role in the generation and maintenance of SSVs or a role in the interaction of vesicles with cytoskeletal elements (Johnston et al., 1989a).

2.2.4 Synaptobrevin

Synaptobrevin is a protein which exists in two highly homologous isoforms which are also referred to as VAMP-1 and VAMP-2 (Trimble et al., 1988; Baumert et al., 1989; Elferink et al., 1989; Südhof et al., 1989b). It is an abundant integral membrane protein of SSVs with an apparent molecular weight of 18,000 daltons when separated by SDS-polyacrylamide gel electrophoresis (Fig. 2-1). The protein was cloned and sequenced from several species (Trimble et al., 1988; Südhof et al., 1989b; Elferink et al., 1989) and monoclonal and polyclonal antibodies are available for its characterization (Baumert et al., 1989).

Synaptobrevin is a small protein, consisting of 116 amino acids in mammals, with no signal sequence. Most of the amino acid residues are highly hydrophilic, with only one predicted membrane spanning domain close to the C-terminal end of the molecule (Trimble et al., 1988; Südhof et al., 1989b; Elferink et al., 1989; Fig. 2-6). Limited proteolysis of synaptobrevin reconstituted in proteoliposomes provided support for the view that the protein spans the membrane in its entirety (Südhof et al., 1989b).

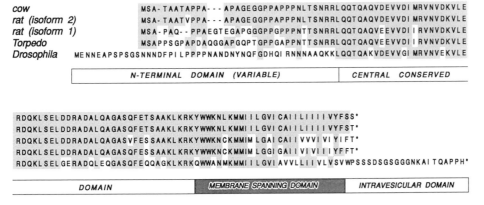

Fig. 2-6 Sequence comparison of synaptobrevin from different species. Amino acids are given in single letter code. Adapted from Südhof et al., 1989b. The sequences for *Torpedo* and rat are from Trimble et al. (1988) and Elferink et al. (1989), respectively.

The striking feature of synaptobrevin is its high degree of conservation in phylogenetically distant animal species. Synaptobrevin is the first vesicle protein identified in invertebrates (*Drosophila*, see Südhof et al., 1989b). A comparison of the sequences published so far (Fig. 2-6) reveals four domains (Südhof et al., 1989b). The N-terminal region is of variable size and is divergent in distant species, with no significant homologies between *Drosophila* and vertebrates. This region also contains the major differences between the two isoforms found in the rat and ends in dibasic residues. It is followed by a stretch of 63 amino acids which is conserved to an extraordinary degree. Approximately 80% of the residues are identical between *Drosophila* and cow. In addition, the few changes are mostly due to replacements by amino acids with very similar properties (Glu for Asp and Ile for Met). The degree of conservation is much higher than that observed for synaptophysin between *Torpedo* and mammals (see above), indicating an essential functional role of this domain. It ends in a cluster of four positively charged amino acids. This part is followed by the membrane-spanning domain which consists of 20 amino acids, also well conserved, with a high average hydrophobicity. The C-terminus, which probably extends into the lumen of the vesicle, is again divergent. It is very short in rat, cow and *Torpedo* (2 amino acids) but comparatively long in *Drosophila* (22 amino acids) (Trimble et al., 1988; Südhof et al., 1989b, Elferink et al., 1989; see Fig. 2-4).

At the present time, little information is available about the biochemical properties of synaptobrevin. It is tempting to speculate, however, that synaptobrevin has an essential function in the exo-endocytotic cycle of SSVs. In contrast to synaptophysin, the most conserved part of the synaptobrevin molecule is located in the cytoplasmic domain directly adjacent to the surface of the vesicle suggesting that its main function resides in this domain. Clearly, synaptobrevin is a strong candidate for mediating membrane interactions involved in exo-endocytosis, but its function remains to be resolved.

2.2.5 P65

P65 was the first of the major integral SSV membrane proteins to be discovered (Matthew et al., 1981). It was characterized with the aid of two monoclonal antibodies (48 and 30) raised against rat brain membranes. P65 appears as a single, somewhat "fuzzy" band of 65,000 daltons when analyzed by SDS-polyacrylamide electrophoresis (Fig. 2-1). A similar, probably identical protein was also independently described (Obata et al., 1987). The protein appears to contain conserved domains since the monoclonal antibodies cross-react with similar proteins in lower vertebrate species (Matthew et al., 1981). The protein was purified from rat brain by affinity chromatography but was not further characterized (Bixby and Reichardt, 1985).

Again, the function of this protein is not known. However, two interesting properties were described which might offer some clues to the functional role of p65. A calmodulin-binding protein was identified in secretory vesicles from several organs (Bader et al., 1985; Fournier and Trifaro, 1988) which was later identified as p65 (Trifaro et al., 1989). If these results will be confirmed by future studies, p65 would represent the major calmodulin-binding protein of synaptic vesicles, making it a second candidate, in addition to synaptophysin, for transducing the calcium signal in exocytosis. In addition, it was found recently that a vesicle protein which is probably identical to p65, possesses agglutinating activity and binds gangliosides (Popoli and Mengano, 1988; Popoli et al., 1989). The primary structure of p65 has recently been determined (Perin et al., 1990). It is a transmembrane glycoprotein of 421 amino acid residues with one membrane spanning domain close to the N-terminus (the latter is located inside the vesicle). The cytoplasmic tail of p65 contains two copies of a 116 amino acid repeat which share 41% identical amino acid residues. These repeats are separated from the membrane by a doamin of highly charged residues. The cytoplasmic repeats exhibit striking homology to the regulatory (C_2) domain of protein kinase C (identity between 32 and 40% depending on the isoform). A p65 expression construct covering the two repeats, resulted in a recombinant protein which binds acidic phospholipids with high affinity and is capable of agglutinating red blood cels. Preliminary evidence indicated that the recombinant protein binds to the hydrophobic portion of the phospholipids and may insert partially into the bilayer (Perin et al., 1990). Thus, p65 has emerged as a strong candidate for mediating Ca^{++}-binding and membrane fusion during exocytosis.

2.2.6 SV2

SV2 was discovered when monoclonal antibodies were raised against vesicle proteins of the marine electric ray and was subsequently found to be present also in mammalian synaptic vesicles (Buckley and Kelly, 1985). A similar, probably identical protein, was characterized independently (Walker et al., 1986).

At the present time, structural information of SV2 is limited. The apparent molecular weight of SV2 is approximately 100,000 daltons. Pulse-chase experiments in PC12 cells revealed that a precursor form of 64,000 daltons exists which is immunoprecipitated by the monoclonal antibody. These results indicate that the protein is highly glycosylated (Buckley and Kelly, 1985).

Recently, the isolation of cDNA clones was reported, increasing the possibility that the primary structure will be determined in the near future (Buckley and Kelly, 1989).

2.2.7 P29

P29 is an integral membrane protein with an apparent molecular weight of 29,000 daltons and an isoelectric point of approximately 5.0 (Baumert et al., 1990). The protein is not glycosylated. Apparently, it is less abundant than synaptophysin, synapsin I, p65 or synaptobrevin (see Fig. 2-1) but the precise amount of p29 in purified vesicles remains to be determined (Baumert et al., 1990).

P29 exhibits two interesting properties. First, two monoclonal antibodies raised against p29 display a limited cross-reactivity with synaptophysin, indicating the presence of related epitopes. Second, p29, like synaptophysin, is phosphorylated on tyrosine residues by endogenous tyrosine kinase activity (Baumert et al., 1990; see also Pang et al., 1988a). It will be of interest to evaluate the precise relationship between these two proteins when the amino acid sequence of p29 is elucidated.

2.2.8 Rab3

Rab3 is a GTP-binding protein (G protein) belonging to the class of the so-called small, ras-related G proteins. This class includes more than 20 monomeric proteins with a molecular weight in the range of 20,000 – 28,000 daltons. They are in part membrane-bound but do not contain membrane-spanning segments in their amino acid sequence (Touchot et al., 1987; Barbacid, 1987; Chardin, 1988). As for many other small G-proteins, rab3 was first identified by cDNA cloning (Touchot et al., 1987) and shown to be specific for neurons and endocrine cells by Northern blotting (Matsui et al., 1988; Olofsson et al., 1988; Sano et al., 1989) and immunocytochemistry (Mizoguchi et al., 1989). In a recent study, it was shown that membrane-bound rab3 is exclusively associated with SSVs (Fischer v. Mollard et al., 1990). In addition, a soluble pool of this protein was found to be present in mammalian brain. The nature of the membrane association of rab3 remains to be established but appears to involve a covalent modification. Both the soluble and the membrane-bound form of rab3 were shown to bind detergent (in contrast to rab3 expressed in bacteria), suggesting posttranslational addition of hydrophobic side groups to both forms. Preliminary evidence indicates that the modification of the membrane-bound form may be different from that of the soluble form. In addition, these modifications appear to be different from those of ras proteins which were shown to be farnesylated and partially palmitoylated (Hancock et al., 1989; Fischer v. Mollard et al., 1990).

The function of rab3 is unknown. However, there is increasing evidence that G proteins are involved in directing intracellular membrane interactions, including exocytosis (Howell et al., 1987; Vallar et al., 1987; Bourne, 1988). In particular, the small G proteins sec4 and ypt1 have been implicated in controlling vectorial membrane traffic within the Golgi apparatus and between the Golgi apparatus and the plasma membrane, respectively (Segev et al., 1988; Schmitt et al., 1988; Salminen and Novick, 1987). The existence of free and membrane-bound pools of these proteins led to the suggestion that the proteins cycle between membrane-bound and free states in parallel with the life cycles of the transport vesicles (Segev et al., 1988; Bourne, 1988; Walworth et al., 1989). It is not known whether rab3 undergoes similar cycles and how the specificity of its interaction with SSVs is mediated. However, it is likely that rab3 is involved in controlling a step in the synaptic cycling of SSVs.

2.2.9 Proton ATPase

In contrast to the other SSV-proteins described here, the vesicular proton pump is characterized both with respect to its structure and its function. A full discussion of this enzyme, also referred to as vacuolar or V-ATPase, is beyond the scope of this chapter and the reader is referred to several recent reviews (Rudnick, 1986; Al-Awqati, 1986; Mellman et al., 1986; Schneider, 1987; Scharschmidt and Van Dyke, 1987; Nelson and Taiz, 1989; Forgac, 1989).

Soon after the development of vesicle purification procedures it was reported that synaptic vesicles contain ATPase activities which differ from the sodium-potassium ATPase (Hosie, 1965; Breer et al., 1977). However, only recently it became clear that most of this ATPase activity is contributed by a proton pump (Stadler and Tsukita, 1984; Harlos et al., 1984; Hell et al., 1988) which belongs to the vacuolar class of enzymes (Hell et al., 1988; Wang et al., 1988; Cidon and Sihra, 1989; Yamagata and Parsons, 1989; but see also Yamagata et al., 1989, for the characterization of an E_1-E_2 ATPase from *Torpedo* synaptic vesicles). This proton pump generates an electrochemical potential which provides the energy for the uptake of neurotransmitters (Marshall and Parsons, 1987; Philippu and Matthaei, 1988; Maycox et al., 1990). The class of vacuolar proton pumps is primarily responsible for the acidification of intracellular organelles (yeast and plant vacuoles, lysosomes, endosomes, coated vesicles, secretory organelles etc.; for review see Rudnick, 1986; Forgac, 1989). In certain cases, it is incorporated into the plasma membrane, e.g. in kidney epithelia (Brown et al., 1988) and may play a role in acid regulation (Al-Awqati, 1986; Forgac, 1989). Thus, this enzyme is not specifically located on SSVs but is shared with a wide range of intracellular membranes which provide a functional interconnection between the Golgi complex and the plasmalemma (Rudnick, 1986; Forgac, 1989).

The structural characterization of the vacuolar proton pump was the subject of intense effort of several different laboratories. Most of its subunits have been cloned and sequenced from different sources (Mandel et al., 1988; Manolson et al., 1988; Hirsch et al., 1988; Zimniak et al., 1988; Bowman et al., 1988a; Bowmann et al., 1988b; Nelson et al., 1989; Südhof et al., 1989c). There is general agreement that the pump is a large hetero-oligomer of approximately 300–700 kDa. However, the precise subunit composition is not known and may vary in different organelles or different species, those from plants and lower eukaryotes having a simpler subunit structure (for an overview see Forgac, 1989). Several subunits display limited sequence homologies to corresponding subunits of the mitochondrial F_0-F_1-ATPase indicating that they evolved from a common ancestral gene. The pump contains a strongly hydrophobic subunit which is covalently modified by DCCD and has been shown to form the proton channel (Sun et al., 1987). Furthermore, the ATP-binding site was localized on the 70 kDa subunit (for review see Forgac, 1989).

In addition to the proton pump, a chloride channel was found in synaptic vesicles which provides compensatory chloride influx during vesicular acidification (Maycox et al., 1988; Jahn et al., 1990). The channel protein has not been identified. So far, it is not known if and how this channel is related to chloride channels characterized in other endomembrane systems, e.g. endosomes (Mellman et al., 1986; Schmid et al., 1989).

2.2.10 Additional Proteins Associated with SSV Membranes

2.2.10.1 Proteoglycan

A proteoglycan originally identified in synaptic vesicles of *Torpedo* (Stadler and Dowe, 1982; Walker et al., 1983; Carlson and Kelly, 1983) was also found to be present in mammalian synaptic vesicles. It is highly antigenic and represents the major antigen of most of the early antibodies raised against synaptic vesicles of electric rays (e.g. Hooper et al., 1980; Carlson and Kelly, 1980; Von Wedel et al., 1981; Theresa-Jones et al., 1982). The molecule is strongly acidic, being rich in glucosamine, uronic acid and sulfate (Carlson and Kelly, 1983). The proteglycan apparently lacks a cytoplasmic domain since it is inaccessible to antibodies or protein labeling agents unless the vesicles are disrupted (Carlson and Kelly, 1983). Thus, it appears to be located exclusively in the interior of the vesicle, being partly soluble. The proteoglycan accounts for a major part of the total protein of cholinergic vesicles. It is found only on a subset of synapses in the mammalian nervous system. In the *Torpedo* electric organ, part of the proteoglycan appears to be secreted during exocytosis since recycling vesicles contain much less of this constituent (Stadler and Kiene, 1987). However, in another study only a small amount of it is released from *Torpedo* synaptosomes upon stimulation (Kuhn et al., 1988). This discrepancy needs to be resolved. An immunologically related, but structurally slightly different proteoglycan was found within the extracellular matrix (Caroni et al., 1985; Carlson et al., 1986; Stadler and Kiene, 1987). It is not clear whether this matrix proteoglycan is derived from synaptic vesicles. A better understanding of the role of these molecules in the synaptic membrane life cycle will be dependent on the development of improved tools for the study of their protein core.

2.2.10.2 Synaptic Vesicle Proteins from the Electric Organ of Electric Rays

As mentioned above, synaptic vesicles of the electric organ of electric rays can be isolated in a highly purified preparation and have been studied intensely. It would be expected that proteins essential for vesicle function are conserved from *Torpedo* (or *Narcine*) to mammals. Major proteins first discovered in *Torpedo* (SV2, proteoglycan, synaptobrevin) were subsequently discovered in mammals and, vice versa, mammalian proteins were discovered in *Torpedo* synaptic vesicles, e.g. synaptophysin (Cowan et al., 1990) and synapsin I (Volknandt et al., 1987). Some proteins, however, were only found in *Torpedo* and the occurrence of corresponding proteins in mammals remains to be demonstrated. These include the ATP-transporter which was identified by different labeling strategies as a 34,000 dalton protein (Stadler and Fenwick, 1983; Lee and Witzemann, 1983; Lee and Witzemann, 1987) and two more proteins of unknown function: VAT-1 which is tissue-specific for electromotor neurons (Linial et al., 1989), and SVP 25, an integral membrane glycoprotein (Volknandt et al., 1989).

2.3 Membrane Proteins of Large Dense Core Vesicles (LDCVs)

2.3.1 Preliminary Remarks

In contrast to SSVs, which can be isolated in high yield and purity, LDCVs can only be prepared in minute quantities, partially contaminated with other organelles. Therefore, relatively little information is available about their membrane composition. LDCVs were isolated in sufficient quantity and purity only from highly specialized organs. Relatively pure LDCV fractions were obtained from the neurohypophysis (Nordmann et al., 1979; Gratzl et al., 1980; Russell, 1980;). In addition, secretory granule fractions can be obtained from the adrenal medulla (Smith and Winkler, 1967). These granules are thought to be closely related to LDCVs of catecholaminergic neurons. Furthermore, preparations highly enriched in LDCVs of sympathetic nerves were obtained from splenic nerve (Klein et al., 1982). A second problem associated with the study of LDCV membranes is that they are filled with a heterogenous cocktail of proteins (see Chapter 3) which must be differentiated from the membrane proteins. It is therefore not surprising that most of the identified membrane proteins of LDCVs are enzymes or transporters for which functional assays are available such as those associated with uptake, biosynthesis or processing of catecholamines and neuropeptides. Thus, most of the information available for these proteins is derived from work on specific systems (e.g. sympathetic neurons, adrenal medulla, anterior and posterior pituitary) and little is known about their distribution in the central nervous system. At the present time, no general membrane marker is available for neuronal LDCVs. However, the recent advances in elucidating the primary structure of some of these proteins should aid in the development of improved tools (cDNA-probes, antibodies) for anatomists and cell biologists.

2.3.2 Membrane Proteins Involved in Catecholamine Uptake and Hydroxylation

It is well established that noradrenaline is synthesised from dopamine within the lumen of catecholaminergic vesicles. This reaction is mediated by dopamine -β-hydroxylase. In addition, the following components are involved: a transport system for monoamines, an energizing system for the uptake which is provided by a proton pump of the vacuolar type (see section 3.2.9), and a transmembrane shuttle for electrons which is provided by cytochrome b_{561}. All of these proteins are well characterized (for review see Winkler et al., 1986; Njus et al., 1986; Johnson, 1988; Philippu and Matthaei, 1988), and the primary structures of dopamine-β-hydroxylase (Lamouroux et al., 1987), cytochrome b_{561} (Perin et al., 1988) and of most subunits of the proton pump (see section 2.2.9) were recently elucidated. Dopamine-β-hydroxylase is a dimeric protein, consisting of two identical subunits of 65,000 daltons. Aside from the signal sequence, no obvious membrane spanning region was found in the amino acid sequence (Lamouroux et al., 1987). The biochemical nature of its membrane binding is still unclear. Apparently, no part of the molecule extends to the cytoplasmic side of the vesicle membrane (Taylor et al., 1989). The monoamine carrier has recently been identified by affinity-labeling (Isambert and Henry, 1985; Isambert et al., 1989). It appears to be a membrane protein of approximately 70,000 daltons which was confirmed by radiation-inactivation analysis (Gasnier et al., 1987). Antibodies or structural data are not yet available.

Cytochrome b_{561} is a hydrophobic protein, with 6 membrane spanning domains and short cytoplasmic tails at both the N- and C- terminal ends of the molecule. The protein lacks a signal sequence and is not N-glycosylated (Perin et al., 1988). It is still unclear how the heme groups are coordinated in the molecule (Esposti et al., 1989). This protein is also involved in the processing of neuroactive peptides stored within LDCVs and secretory granules since it provides electrons required for the catalytic cycle of the peptidyl-glycine α-amidating monooxygenase (Kent and Fleming, 1987; Pruss and Shepard, 1987). Although a recent study suggests a widespread distribution of this protein in neurons and neuroendocrine cells (Russell, 1987; Pruss, 1987; Pruss and Shepard, 1987), a thorough immunocytochemical study investigating its tissue and subcellular distribution is still required.

It should be noted that all of these proteins may also be present in SSVs of the sympathetic nervous system (reviewed in Winkler, 1988; Thureson-Klein and Klein, 1990).

2.3.3 Membrane Proteins Involved in Peptide Processing

The last steps in propeptide-peptide conversion occur during maturation of the secretory vesicle. These include proteolytic trimming by highly specific exo- and endopeptidases, acetylation and amidation (Russell, 1987; Loh, 1987). Various enzymes present in peptide containing secretory vesicles have been purified, and the primary structure of some of them, including the amidating enzyme (peptidyl-glycine α-amidating monooxygenase) (Eipper et al., 1987; Stoffers et al., 1989) and carboxypeptidase E (Fricker et al., 1986) have been reported (see Fisher and Scheller, 1988; Fuller et al., 1988; Fricker, 1988; Eipper and Mains, 1988; Sossin et al., 1989; Mains et al., 1990 for review). Although these enzymes appear to be mostly soluble, it appears that some of them may be synthesized as membrane-bound precursors, containing a membrane-spanning domain close to their C-terminus (e.g. peptidyl-glycine α-amidating monooxygenase; Eipper et al., 1987; or carboxypeptidase E; Fricker et al., 1986). A similar domain was found in KEX2, an endopeptidase studied in yeast, that is able to cleave correctly prepro-opiomelanocortin when introduced into a mammalian cell line (Thomas et al., 1988b). The mammalian counterpart, however, has not been identified.

These proteins are present in a variety of peptide secreting endocrine cells and in the nervous system (Russell, 1987; Fricker, 1988; Eipper and Mains, 1988), making them candidates for general markers specific for LDCV membranes. However, very little is known about their cellular and subcellular distribution in the mammalian CNS since most of them are present in low amounts.

2.3.4 Membrane Proteins of Unknown Function

For the reasons outlined above, little is known about LDCV membrane proteins which cannot be identified by functional assays. A series of membrane-bound glycoproteins was described from the chromaffin granule (for review see Winkler et al., 1986). One of them, termed GPII, has a widespread distribution in endocrine and exocrine glands, including the posterior pituitary (Obendorf et al., 1988a). It exhibits a molecular weight of approximately 100,000 daltons. It is different from the SSV-protein SV2 (H. Winkler, personal communication) but appears to be similar, if not identical, to a glycoprotein first characterized in secretory granules of the exocrine pancreas, termed GP-2 (Obendorf et al., 1988a). The pancreatic GP-2 is processed from a membrane-bound to a soluble form during granule maturation (Havinga et al., 1984; Havinga et al., 1985; Paquette et al., 1986). Recently, it was shown that it is not an integral membrane protein but anchored via covalent bonds to phosphatidylinositol (Le Bel and Beattie, 1988). It remains to be established whether GPII or one of the other glycoproteins of the chromaffin granule membrane is also found in LDCVs of the brain.

A different protein with an apparent molecular weight of 120,000 daltons was recently characterized from the membrane of neurosecretory granules of the posterior pituitary (Caorsi and Gonzalez, 1989). Immunocytochemistry revealed a specific localization in the hypothalamo-hypophyseal system, indicating that it may not occur on LDCVs of other neurons.

2.4 Cellular and Subcellular Distribution of SSV Proteins

In the following paragraphs, we will attempt to give a short overview about the cellular and subcellular localization of synaptic vesicle proteins, with special reference to work done on SSV proteins. This discussion is largely restricted to mammalian species.

2.4.1 Distribution of SSV Proteins in the Nervous System

2.4.1.1 Cellular and Tissue Distribution

All proteins which in neurons are specific for SSV membranes, namely synapsin I and II (De Camilli et al., 1983a; Südhof et al., 1989a), synaptophysin (Wiedenmann and Franke, 1985; Navone et al., 1986), p29 (Baumert et al., 1990), synaptobrevin (Baumert et al., 1989), p65 (Matthew et al., 1981) and SV2 (Buckley and Kelly, 1985), are widely distributed in nerve terminals of both the central and peripheral nervous system (Figs. 2-7–2-9). Light microscopy immunostaining of a variety of tissues and brain regions with antibodies directed against any of these proteins produced similar staining patterns (Fig. 2-9) corresponding to those expected for nerve terminals. Typically, isolated puncta or rows of puncta corresponding to "en passant " varicosities were observed, each punctum representing a single nerve terminal (Figs. 2-7–2-9). In some cases, staining patterns for different SSV proteins were virtually identical. In other cases, differences in the staining intensity of individual nerve terminals suggested some variation in the stoichiometric ratio of the various proteins between different nerve terminals. For some of these proteins, notably synapsin and synaptobrevin, this may be explained partially by the existence of different molecular isoforms. Antibodies specific for each of the four synapsins revealed that the relative concentration of the four molecules varies in individual nerve terminals. Some terminals appear to lack certain isoforms (Südhof et al., 1989a). Similarly, the two known isoforms of synaptobrevin (VAMP-1 and VAMP-2) have a different, although partially overlapping, distribution (Elferink et al., 1989). So far, no clear correlation between this heterogeneity and other properties of nerve endings, such as the neurotransmitter phenotype, has been found.

Fig. 2-7 Localization of synaptophysin by immunofluorescence in a 1μm-thick plastic section of the rat cerebellar cortex. The staining pattern is represented by puncta of immunoreactivity (white) which represent individual nerve endings. In the molecular layer (ML) they are densely apposed to each other and outline the dendritic trees(d) of Purkinje cells (PC). Small arrows point to axo-somatic synapses on Purkinje cell perikarya and a long arrow to the faint immunoreactivity in the Golgi complex of one of these cells. Clusters of puncta in the granule cell layer (GL) represent synaptic glomeruli (g). Calibration bar: 30 μm.

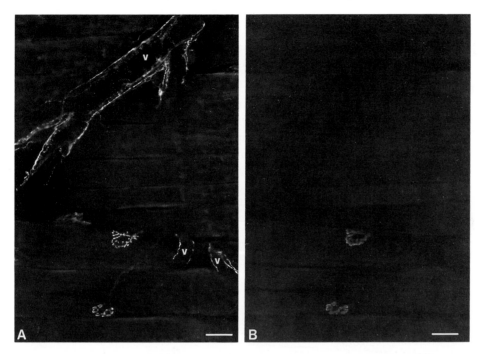

Fig. 2-8 Localization of synapsin I by immunofluoresence in the skeletal muscle of the rat (extensor digitorum longus). Synapsin I immunoreactivity (demonstrated by rhodamine fluorescence) is shown in (A). Both varicose nerve endings, which surround and outline blood vessels (v), and motor nerve terminals are immunoreactive. (B) shows the same field shown in (A) after counterstaining with fluorescein-conjugated α-bungarotoxin (a toxin which binds to the nicotinic cholinergic receptor) to reveal the location and morphology of motor end plates. Calibration bars: 20 μm.

SSV proteins were also detected at unexpected locations. For example, some of these proteins were found in a variety of sensory nerve endings including muscle spindles, Golgi tendon organs, and sensory endings of the inner ear (De Camilli et al., 1988a; Scarfone et al., 1988; Baumert et al., 1990). 50 nm vesicles, reminiscent of SSVs, have been observed in these sensory nerve endings by electron microscopy, but a secretory function has not been documented. In addition, these proteins are present in nerve terminals of the posterior pituitary whose only known function is to secrete neuroactive peptides into blood capillaries via a class of LDCVs, the so called neurosecretory granules. In these terminals, SSV proteins were found to be selectively localized on vesicles morphologically identical to SSVs (Navone et al., 1989). The function of these microvesicles, which are clustered at the side of the ending directly facing blood capillaries, is unknown (for review see Morris et al., 1987).

Conversely, synapsin I and synapsin II are lacking from a specific group of presynaptic elements, those forming the so-called ribbon synapses, although other SSV proteins are present in these terminals (Favre et al., 1986; Scarfone et al., 1988; Mandell et al., 1989). Ribbon synapses are found in sensory organs and possess special morphological and functional properties. The presynaptic elements are located either in the cell body of sensory cells (e.g. hair cell of the inner ear) or at the end of small cellular processes very different from typical axons (e.g. photoreceptors, retinal bipolar cells). They contain a peculiar electron-dense structure, the "ribbon" or ribbon-like structure which is surrounded by SSVs (Sjöstrand, 1953; Smith and Sjöstrand, 1961; Kidd, 1962; Flock, 1964). The ribbon, which is always directly adjacent to postsynaptic sites, is thought to be involved in directing SSVs to release sites (Bunt, 1971; Gray and Pease, 1971). In ribbon-containing nerve terminals, transmitter release is controlled by graded potentials rather than by action potentials (Roberts and Bush, 1981). It is possible that the lack of all synapsins from these synapses is

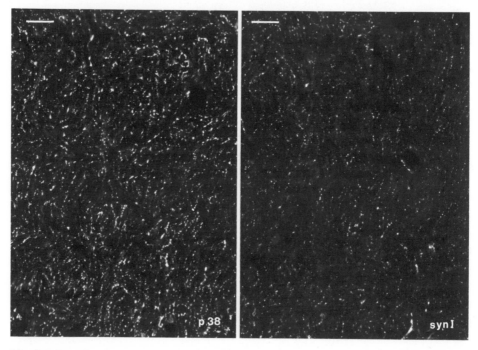

Fig. 2-9 Localization of synaptophysin (p38) and synapsin I (syn I) by immunofluorescence in the muscular layer of the vas deferens of the rat. Rows of small fluorescent puncta represent varicose nerve terminals interspersed with layers of smooth muscle cells. Calibration bar: 50 μm. Reproduced from Navone et al. (1988) by copyright permission of Munksgaard International Publishers.

related to these highly peculiar properties, reflecting a difference in the molecular mechanisms of SSV clustering and exocytosis.

2.4.1.2 Subcellular Localization

The widespread presence of synapsin (De Camilli et al., 1983a; De Camilli et al., 1983b; Navone et al., 1984; Südhof et al., 1989a; Navone et al., 1989),

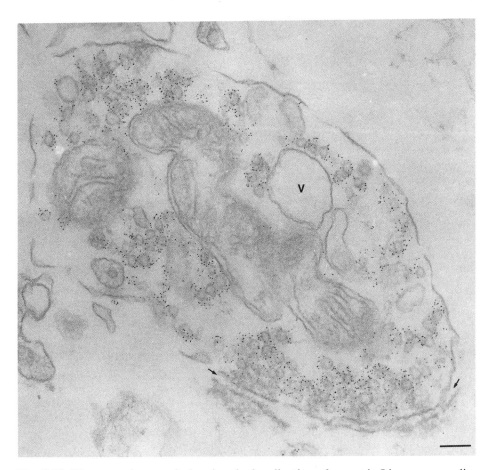

Fig. 2-10 Electron micrograph showing the localization of synapsin I in a nerve ending of the bovine hypothalamus. Immunogold labeling of a lysed isolated nerve ending. Gold particles are selectively localized on SSVs. Note lack of labeling of mitochondria, of large vacuolar profiles (v), and of the plasmalemma, including its presynaptic portion (portion between two arrows) recognizable by the clustering of SSVs and by the presence of an attached piece of post-synaptic spacialization. Calibration bar: 140 nm. Reproduced from De Camilli et al. (1988b) by copyright permission of Alan R. Liss Publishers.

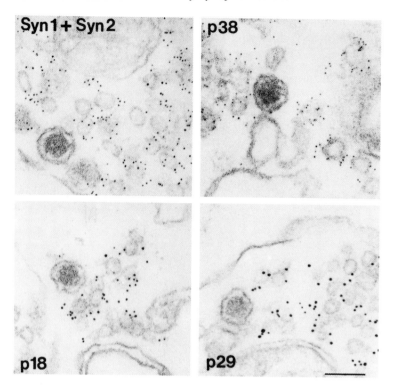

Fig. 2-11 Selective localization of synapsin, synaptophysin (p38), p18 and p29 on SSVs in nerve terminals of the bovine hypothalamus. Electron micrographs showing immunogold labeling of lysed synaptosomes. In all fields gold particles are concentrated on the surface of SSVs.Virtually no gold labeling is visible on the surface of LDCVs or on other membranes. Calibration bar: 100 nm. Reproduced from Navone et al.(1986) and Baumert et al. (1990) by copyright permission of the Rockefeller University Press.

synaptophysin (Wiedenmann and Franke, 1985; Navone et al., 1986; Tixier-Vidal et al., 1988), synaptobrevin (Baumert et al., 1989), p65 (Matthew et al., 1981; Navone et al., 1989), p29 (Baumert et al., 1990) and SV2 (Buckley and Kelly, 1985) on virtually all SSVs was documented by a variety of electron microscopic techniques including labeling of ultrathin frozen sections, pre-embedding immunoperoxidase, and immunogold and immunoferritin labeling of brain tissue, synaptosomes, or purified vesicle fractions (Figs. 2-10 and 2-11; see also Jahn and Maycox, 1988; Hell et al., 1988; Baumert et al., 1989). Further confirmation was obtained by subcellular fractionation of rat brain. Synapsin I and II, synaptophysin, synaptobrevin, p65 and p29 copurify with synaptic vesicles (Huttner et al., 1983; Jahn et al., 1985; Browning et al., 1987; Baumert et al., 1989; Floor and Feist, 1989; Baumert et al., 1990; and unpublished observations) although the idea that synaptophysin and p65 are colocalized has recently been disputed (Volknandt et al., 1988).

SSVs are organelles which are functionally interconnected by membrane fusion-fission events with other intracellular membranes such as the plasma membrane, endocytic vesicles and vesicles of the trans Golgi network (for review see Kelly, 1985; Mellman et al., 1987; Burgess and Kelly, 1987; De Camilli and Jahn, 1990). Accordingly, small pools of these proteins have been detected in cisternae in nerve teminals, tubulo-vesicular elements and multi-vesicular bodies in axons and pleomorphic vesicles in the trans region of the Golgi complex (Navone et al., 1986; Baumert et al., 1989; Tixier-Vidal et al., 1988; Baumert et al., 1990; see Janetzko et al., 1989, for the Golgi-localization of SV2 in *Torpedo* neurons). In this respect, some differences were observed between the proteins synaptophysin, p29 and p65 and the proteins synapto-brevin and synapsin I. The former proteins were clearly present in the region of the Golgi complex where they are probably localized in newly synthesized vesicles or in vesicles recycled from nerve terminals (see section 2.5.2.1). Synapsin I and synaptobrevin, however, were not detectable in the Golgi complex region (Navone et al., 1986; Baumert et al., 1989). The synapsins, as peripheral membrane proteins (see section 2.2.2), are synthesized on free ribosomes in the cytoplasm and associate posttranslationally with their target membranes. In contrast, synaptobrevin is thought to span the entire vesicle membrane (see section 2.2.4), but it is not clear whether it is synthesized on free or membrane-bound ribosomes. Lack of detectable immunoreactivity for these proteins in the region of the Golgi complex raises the possibility that the association of both proteins with SSV membranes occurs distally to the Golgi complex.

It has been shown that expression of several of these proteins during development increases in parallel with synaptogenesis (see e.g. Lohmann et al., 1978; Greif and Reichardt, 1982; De Camilli et al., 1983a; DeGennaro et al., 1983; Levitt et al., 1984; Knaus et al., 1986; Weiss et al., 1986; Chun and Shatz, 1988; Leclerc et al., 1989). A thorough study of changes in the pattern of intracellular distribution of synaptophysin and synapsin I during development has been recently carried out in cultured hippocampal neurons (Lindsley et al., 1987; Fletcher et al., 1989; Fletcher, De Camilli and Banker, manuscript in preparation). These cells express fully differentiated axonal and dendritic processes *in vitro* (Banker and Waxman, 1988). Both synapsin I and synapto-physin are preferentially targeted to axons from the beginning of their differentiation. However, they exhibit a rather diffuse distribution in the distal axons until appropriate contacts with postsynaptic cells are made. When such contacts are established, both proteins undergo rapid redistribution and cluster at presynaptic contact sites. This suggests that SSVs or their respective precursor membranes are concentrated in distal axons before synapses are formed. It remains to be seen whether synapsin is already associated with the membrane of SSV precursors before they cluster at presynaptic sites. A similar sequence of events has also been observed *in situ* by immunocytochemistry. Non-synaptic accumulations of synapsin I in distal axons were observed at sites where axon endings arrive before functional synapses are established (Chun and Shatz, 1988). Lower levels of synapsin I are observed in distal axons for which arrival at the target sites coincides with the beginning of synaptogenesis (De Camilli et al., 1983a; Levitt et al., 1984; Chun and Shatz, 1988).

In sharp contrast to the abundance of synapsin, synaptophysin, synaptobrevin and p29 in the membranes of SSVs, these proteins were not detectable by immunocytochemical techniques in the membranes of LDCVs (Navone et al., 1984; Navone et al., 1986; De Camilli and Navone, 1987; Baumert et al., 1989; Navone et al., 1989; Baumert et al., 1990). This striking difference was confirmed by subcellular fractionation of nerve terminals of the posterior pituitary where an almost complete separation of the SSV-proteins from neurosecretory granules (the LDCVs of the posterior pituitary) was achieved (Navone et al., 1989). These data show that the membrane composition of SSVs and LDCVs is dissimilar, ruling out that SSVs are derived from recycling LDCV membranes as suggested by others (see also section 2.5.2.1).

It should be noted that the lack of significant amounts of the major SSV membrane proteins from LDCV membranes has been disputed by some authors. Most of the data favoring the presence of these proteins on LDCVs is based on subcellular fractionation experiments carried out with sucrose density gradients (see e.g. Agoston and Whittaker, 1989; Obendorf et al., 1989). A bimodal distribution of synaptophysin was observed on these gradients, the dense fraction comigrating with neuropeptides. However, it cannot be ruled out that in these studies the dense pool represents SSVs associated with other membrane structures, e.g. the presynaptic plasma membrane, unbroken nerve terminals or LDCVs. It is well known that in the CNS a major part of the vesicle fraction remains associated with the plasma membrane and migrates towards dense regions of sucrose gradients during subcellular fractionation (Von Schwarzenfeld, 1979). Therefore, immunogold electron microscopy of these fractions is required before a conclusion can be reached. Clearly, it cannot be ruled out that a minor pool of SSV membrane proteins is present on LDCV membranes. Also, it is possible that the presence of a small pool of SSV proteins in dense gradient fractions is due to their presence on immature LDCVs as suggested by images such as that shown in Fig. 2-17F.

No electron microscopic data are available with respect to the distribution of LDCV membrane proteins in the nervous system. There is little doubt that some proteins are shared with SSVs and other membranes of the vacuolar system, e.g. the proton pump and possibly also the chloride channel. It remains to be established whether LDCVs contain membrane proteins which are specific for these organelles. Hopefully, antibodies will soon be available for LDCV membrane proteins, allowing to close this gap in our knowledge concerning the interrelationship of their membranes with those of other intracellular organelles.

In summary, the available data support the concept that all SSVs in all nerve terminals share a set of abundant membrane proteins independent of their specific neurotransmitter content. Despite some minor degree of heterogeneity, for example that due to isoform variability of synapsin and synaptobrevin, it appears that the SSV membrane has a highly defined composition where the concentration of each protein is controlled. This composition must be maintained during recycling. The stoichiometry of the proteins is still unknown,

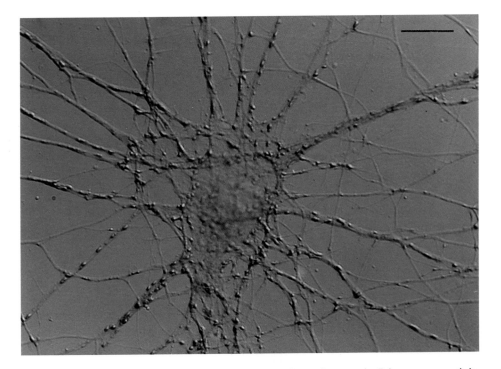

Fig. 2-12 Color micrograph showing the localization of synapsin I immunoreactivity with immunoperoxidase staining in cultured hippocampal neurons. The micrograph was taken with Nomarsky optics to visualize unlabeled cell processes. The perikaryon of a large neuron from which several dendrites irradiate is visible at the center of the field. The meshwork of thin curvilinear processes is represented by axons. Immunoreactivity appears in the form of yellow-brown puncta and rods. As shown by electron microscopy (T. Fletcher, P. De Camilli and G. Banker in preparation), each of them is represented by an axo-dendritic nerve terminal. Calibration bar: 15 μm. Unpublished micrograph by T. Fletcher, G. Banker and P. De Camilli.

but estimates have been made for synapsin I and synaptophysin (for each approx. 20–40 monomer copies/vesicle). It is difficult to imagine how all these proteins can fit onto the vesicle surface (a single IgG molecule, for example, is clearly visible as a major bleb on the vesicle surface in the electron microscope (unpublished observations)). It is tempting to speculate that at least some of the major proteins form a semirigid superstructure which might, in addition, explain the highly uniform size and the remarkable osmotic stability of SSVs. This should become clear in future studies.

2.4.1.3 SSV Proteins as Marker for Synapses

Due to their specific localization in nerve terminals, SSV proteins represent optimal markers for synapses and for nerve terminals in general (Fig. 2-12). Antibodies directed against these proteins can be used to visualize and quantitate synapses in immunocytochemical and immunochemical studies (see e.g. Goelz et al., 1981; De Camilli et al., 1983a; Weiss et al., 1986; Metz et al., 1986; Brock and O'Callaghan, 1987; Wharton et al., 1988; Walaas et al., 1988). Their early expression during development makes them unique tools for the study of synaptogenesis (Levitt et al., 1984; Knaus et al., 1986; Weiss et al., 1986; Burry et al., 1986; Mason 1986; Chun and Shatz, 1988; Leclerc et al., 1989). In addition, synapsin I which is expressed at significant concentration only in neurons, is one of the best neuronal markers available (De Camilli et al., 1983a). Its expression by a given cell can be considered as an important criterion for neuronal differentiation (see section 2.4.2).

2.4.2 Distribution of SSV Proteins Outside the Nervous System: Localization in Synaptic-Like Microvesicles (SLMVs) of Endocrine Cells

Although SSVs are generally thought to be organelles specific for nerve endings, many of their peculiar and abundant membrane proteins, namely synaptophysin, synaptobrevin, p29, p65 and SV2, were detected at high concentrations also in endocrine cells specialized for the regulated secretion of peptide hormones (Matthew et al., 1981; Buckley and Kelly, 1985; Wiedenmann and Franke, 1985; Navone et al., 1986; Baumert et al., 1989; Baumert et al., 1990). These include cells of the anterior pituitary, chromaffin cells of the adrenal medulla, endocrine cells of the pancreas (Fig. 2-13), cells of the diffuse

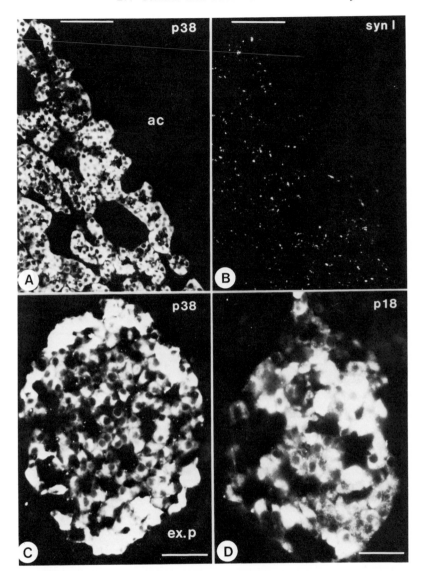

Fig. 2-13 Localization of synaptophysin (p38), synapsin I (syn I) and synaptobrevin (p18) in endocrine tissues by immunofluorescence. (A) and (B) show sections of the adrenal gland, (C) and (D) sections of pancreatic islets. Synaptophysin and synaptobrevin immunoreactivities are present both in nerve terminals (small arrows) and in the cells of the adrenal medulla and endocrine pancreas. Synapsin I immunoreactivity is detectable only in nerve terminals. No immunoreactivity for any of these proteins is visible in cells of the adrenal cortex (ac) or of the exocrine pancreas (ex.p). Calibration bars: (A) and (B), 80 μm; (C) and (D), 40 μm. Reproduced from Navone et al (1986), Baumert et al. (1989) and Baumert et al. (1990) by copyright permission of the Rockefeller University Press and Oxford University Press.

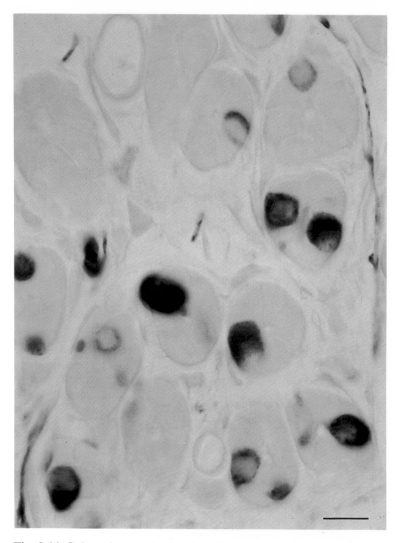

Fig. 2-14 Color micrograph showing the presence of synaptophysin in endocrine cells and axonal varicosities of the stomach mucosa as demonstrated by immunoperoxidase labeling. Calibration bar: 15 μm.

endocrine system of the gut (Fig. 2-14) and C cells of the thyroid. These proteins were also detected in several cell lines derived from such tissues such as GH$_3$ cells, PC12 cells (Navone et al., 1986; Lowe et al., 1988; Baumert et al., in press), insulinoma cells (Fig. 2-15) and AtT20 cells (Lowe et al., 1988; Tooze et al., 1989; own unpublished observations). At least synaptophysin is also present in a variety of tumors derived from neuroendocrine cells (Wiedenmann

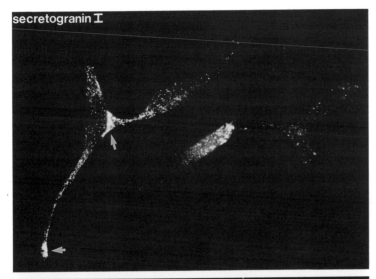

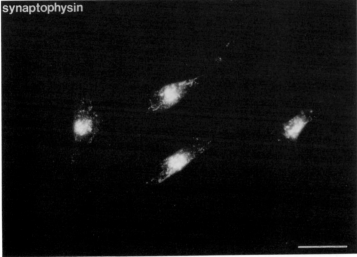

Fig. 2-15 Double-immunofluorescence staining of four insulinoma cells (RIN5) with antibodies against secretogranin I (chromogranin B) and against synaptophysin. The different distribution of the two antigens indicates a localization on different intracellular organelles. Puncta of secretogranin I immunoreactivity, which represent individual secretory granules, are particularly concentrated in distal portions of the cell processes (arrows). Synaptophysin immunoreactivity is particularly concentrated in the centrosomal area. Calibration bar: 25 μm. Unpublished observations by A. Thomas and P. De Camilli: antibodies directed against secretogranin I were kind gift of Drs. Huttner and Rosa.

et al., 1986; Buffa et al., 1987). In fact, synaptophysin (for which polyclonal and monoclonal antibodies are commercially available from several sources) is now regarded as one of the most reliable markers for this class of tumors (Chejfec et al., 1987; Gould et al., 1987; Miettinen, 1987; Bishop et al., 1988; Leube et al., 1988; see also chapter 7). Due to the clinical importance of such markers, it will be of interest to determine whether the other SSV proteins colocalized with synaptophysin in endocrine cells, notably SV2, p65, p29 and synaptobrevin, are also present in these tumors.

In contrast to the integral membrane proteins discussed above, the synapsins are either absent (De Camilli et al., 1978; De Camilli et al., 1983a; Fried et al., 1982; Haycock et al., 1988; Tooze et al., 1989) or present only at very low concentrations (Bähler et al., 1989b) in endocrine tissues. However, they are expressed by endocrine cells grown *in vitro* under conditions which result in the development of neuronal characteristics, i.e. the expression of neurite-like processes (Haycock et al., 1988; Tooze et al., 1989; own unpublished observations). A similar induction of synapsin I expression was observed in PC12 cells which were differentiated by nerve growth factor (Romano et al., 1987). In contrast, the level of synaptophysin expression remained unchanged

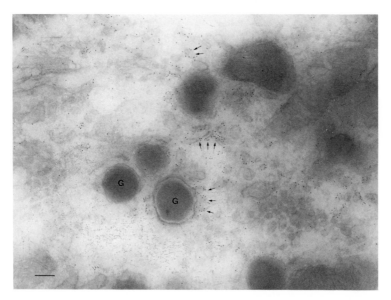

Fig. 2-16 Localization of synaptophysin on microvesicles distinct from secretory granules in a cell of the anterior pituitary. Electron micrograph showing an ultrathin frozen section labeled by immunogold. Gold particles are localized on pleomorphic microvesicles (arrows). Some of them are directly adjacent to granule (G) membranes. Calibration bar: 130 nm. Reproduced from Navone et al., (1986) by copyright permission of the Rockefeller University Press.

(unpublished observations) or rather decreased under these conditions (Vetter and Betz, 1989).

The subcellular distribution of some of the SSV membrane proteins in endocrine cells was analyzed by both light and electron microscopy immuno-

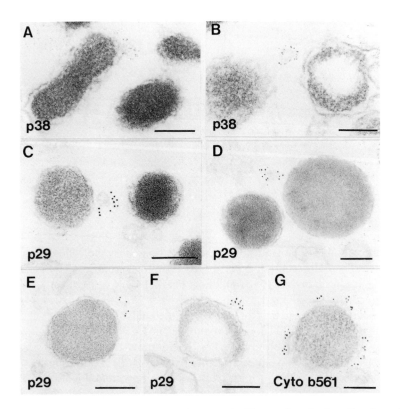

Fig. 2-17 Gallery of electron micrographs illustrating the localization of synaptophysin (p38), p29 and cytochrome b_{561} (cyto b561) in subcellular organelles of endocrine cells. Immunogold labeling of agarose-embedded homogenates of bovine anterior pituitary (A-D) and adrenal medulla (E-G). Both synaptophysin and p29 are selectively localized on microvesicles (synaptic-like microvesicles, SLMVs) and no immunoreactivity for these proteins is visible on secretory granules. In F a p29-positive small membrane evagination of a vacuole with a partially dense core is shown. This structure may represent an immature granule. If so, the figure suggests that separation of SLMV proteins from secretory granule membranes takes place (or is completed) during secretory granule maturation. The intense labeling of secretory granule membranes after cytochrome b_{561} immunolabeling, indicates that their surface is freely accessible to antibodies and represents a positive control for the absence of synaptophysin and p29 immunoreactivity. Calibration bar: 200 nm. Reproduced from Navone et al. (1986) and Baumert et al. (1990) by copyright permission of the Rockefeller University Press.

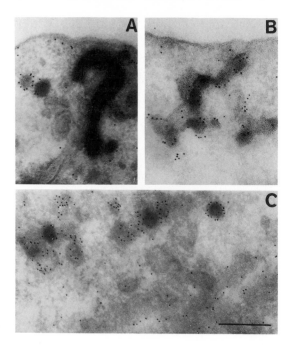

Fig. 2-18 Electron micrograph demonstrating the localization of synaptophysin on small endocytic vesicles in undifferentiated PC12 cells (A and B) and in CHO cells transfected with synaptophysin cDNA (C). Immunogold labeling of cells which have been exposed for 1 hr to an extracellular tracer (horseradish peroxidase) before fixation in order to label endocytic compartments (for experimental details see Johnston et al., 1989b). In both cell types immunogold label selectively decorates the surface of round or elongated microvesicles. Some of them are also positive for the extracellular tracer as shown by their content of dark peroxidase reaction product. Larger peroxidase-positive vacuoles, probably late endosomes or lysosomes, are unlabeled by immunogold. Note lack of labeling of the plasmalemma. Calibration bar: 200 nm. Reproduced from Johnston et al. (1989b) by copyright permission of Oxford University Press.

cytochemistry as well as by subcellular fractionation (Navone et al., 1986; Franke et al., 1986; Obendorf et al., 1988b; Fournier et al., 1989; Baumert et al., 1990). Results from our laboratories suggest that at least synaptophysin, synaptobrevin and p29 are localized on a population of microvesicles clearly distinct from secretory granules. At the light microscopic level of resolution, immunoreactivity for these proteins was concentrated particularly in the region of the Golgi complex while immunoreactivity for content proteins of secretory granules was detected primarily in the peripheral regions of the cell (Navone et al., 1986; see Fig. 2-15). Immunogold electron microscopy showed that both synaptophysin and p29 were found primarily on a population of pleomorphic microvesicles (synaptic-like microvesicles, referred to as SLMVs) which were sparsely distributed throughout the cytoplasm, often present in small groups under the plasmalemma, and concentrated particularly at the trans-side of the Golgi complex and around centrioles. Secretory granules were consistently unlabeled, with the exception of occasional small evaginations of their membranes (Navone et al., 1986; Baumert et al., 1990; Figs. 2-16, 2-17, 2-18a and b). Organelles with such evaginations may represent immature granules. If this is the case, such images may indicate that sorting of SSV membrane proteins from secretory granule membrane proteins is completed during granule maturation (Baumert et al., in press).

Like SSVs, SLMVs are labeled by endocytic tracers, suggesting that they can undergo fusion with the plasmalemma and subsequent reinternalization by endocytosis (Johnston et al., 1989b; Fig. 2-18a and b). In fact, synaptophysin was detected at low concentration in the plasmalemma, and synaptophysin-positive omega figures in the plasmalemma were observed (Navone et al., 1986; Johnston et al., 1989b). SLMVs are likely to be recycling organelles which shuttle between the plasmalemma and the Golgi complex.

The selective localization of synaptophysin and p29 was confirmed by immunoisolation of SLMV membranes from extracts of endocrine cells and tissues using immobilized monoclonal antibodies for synaptophysin as immunosorbent. P29 and synaptobrevin copurify with synaptophysin on the immunobeads (Baumert et al., 1990). In electron microscopy, immunoadsorbed organelles are represented primarily by pleomorphic microvesicles and by only few secretory granules (unpublished observations).

It is still debated whether a low concentration of SSV membrane proteins is also present in mature secretory granule membranes. Again, most of the conflicting data are based on subcellular fractionation, using bovine adrenal medulla or related cell lines as model systems. In these studies, comigration of SSV-markers with those of secretory granules during sucrose density gradient centrifugation was used as criterion for colocalization (Obendorf et al., 1988b; Fournier et al., 1989; Trifaro et al., 1989). In one case, SSV-protein containing organelles were immuno-isolated, resulting in the cosedimentation of 90% of peptide hormone (Lowe et al., 1988). Unfortunately, no immunoelectron microscopy is available for any of these fractions, making it difficult to evaluate whether the secretory granule fractions from the dense parts of the sucrose gradients are contaminated with microvesicles. While most authors presently agree that synaptophysin is primarily concentrated on a population of microvesicles different from secretory granules (Navone et al., 1986; Wiedenmann et al., 1988; Obendorf et al., 1988b; Navone et al., 1989; Baumert et al., 1990) the amount of synaptophysin and of the other SSV proteins present on the secretory granule membrane is a subject of controversy (Schilling and Gratzl, 1988; Wiedenmann et al., 1988; Obendorf et al., 1988b; Fournier et al.,

1989). We have recently re-evaluated this issue using bovine adrenal medulla as model system. While there was a clear codistribution of some synaptophysin with chromaffin granule markers on sucrose density gradients, treatment of the granule fraction with synaptophysin-immunobeads led to an almost complete removal of synaptophysin containing membranes, leaving behind a membrane fraction rich in cytochrome b_{561} and dopamine-β-hydroxylase but with only traces of synaptophysin (Fischer v. Mollard et al., 1990; unpublished observations). These findings suggest that synaptophysin present in chromaffin granule membrane preparations is due mostly to contamination by microvesicles, supporting the view that the membrane composition of SLMVs and secretory granules is dissimilar.

2.5 SSVs and LDCVs – Organelles of Two Distinct Pathways of Regulated Exocytosis in Neurons

The identification and characterization of vesicle-specific membrane proteins, in particular of SSV proteins, has allowed new approaches to the study of the life cycle of these organelles, namely their biogenesis, transport to release sites and exo-endocytosis. In the following sections, we will briefly discuss some of the implications derived from the study of vesicle membrane proteins which are relevant for membrane traffic in neurons and endocrine cells. For a more general overview, the reader is referred to our recent review published elsewhere (De Camilli and Jahn, 1990).

2.5.1 Synthesis and Axonal Transport of SSV and LDCV Membranes

2.5.1.1 SSVs

It is unclear whether SSV membrane proteins leave the Golgi apparatus as part of the mature SSV or in precursor membranes from which SSVs originate at later stages, e.g. after transport into the nerve terminal. As mentioned above, most of the SSV membrane proteins are enriched in the area of the Golgi complex. The morphology of membranes positive for SSV proteins in the vicinity of the Golgi apparatus was recently analyzed in developing hypothalamic neurons (Tixier-Vidal et al., 1988) and in electromotor neurons of *Torpedo marmorata* (Janetzko et al., 1989), using synaptophysin and SV2 as markers, respectively. These membranes are represented by pleomorphic vesicles and vacuoles. However, in these studies newly formed membranes could not be distinguished from retrogradely transported membranes. It is likely (see section 2.4.1.2) that a large fraction of SSV membrane proteins present in the region of the Golgi complex represents membranes recycled from nerve terminals. SV2 containing structures were analyzed in the axons of *Torpedo* electromotor neurons and found to be highly heterogeneous, including SSV-like structures, irregularly shaped vesicles and multivesicular bodies (Janetzko et al., 1989). The first may represent anterogradely transported organelles, the last retrogradely transported organelles. Working with the same system, Kiene and Stadler (1987) used metabolic labeling of the vesicular proteoglycan to study selectively anterograde axonal transport of SSV precursor membranes. According to this study, synaptic vesicles are transported as morphologically homogeneous small vesicular structures which are loaded with acetylcholine only after arrival in the nerve terminal. In contrast, typical 50 nm SSVs did not accumulate proximally to sites where axoplasmic transport was interrupted by a cold block (Tsukita and Ishikawa, 1980).

When motor neurons of the rat sciatic nerve are ligated, equal amounts of SV2 and synaptophysin immunoreactivity accumulate on both sides of the crush whereas synapsin I is found only on the proximal side (Bööj et al., 1986; Dahlström and Bööj, 1988; Bööj et al., 1989). These data are consistent with a steady-state flow of intrinsic SSV membrane proteins in both directions. However, the immunoreactive structures accumulating on both sides of the crush have not yet been identified by electron microscopy. Synapsin I which is transported only distally is probably broken down in nerve terminals.

In conclusion, the nature of the membranous organelles which deliver SSV membrane material to the synaptic terminals is still unclear. At the present time, it cannot be ruled out that SSV membrane proteins are transported as part of heterogeneous tubulo-vesicular elements. These organelles may be closely related to SLMVs of endocrine cells, being converted to homogeneous 50 nm SSVs by a nerve terminal specific mechanism after arrival in nerve endings.

2.5.1.2 LDCVs

Like endocrine secretory granules, LDCVs are assembled in the region of the trans-Golgi network. Aggregation and precipitation of secretory peptides is thought to play a major role in the biogenesis of LDCVs (chapter 3). It has been shown that a given neuron may contain LDCVs with different peptide content, and that different LDCVs may be targeted to distinct cellular compartments. Content heterogeneity of LDCVs is thought to result from a differential insolubilization of different peptide products at the exit from the trans-Golgi network. LDCVs leave the region of the Golgi complex as fully assembled organelles, although content maturation continues during their transport to the sites of release. Axon ligation results in the accumulation of LDCVs proximally to the ligation site (Tomlinson, 1975). The fate of LDCV membranes after exocytosis remains unknown due to the lack of immunocytochemical membrane markers and is still a controversial issue (see section 2.5.2.1).

2.5.2 Exo- Endocytosis of SSVs and LDCVs

2.5.2.1 Membrane Recycling

The recycling of SSVs in nerve terminals has been intensely studied in a number of model systems (Heuser and Reese, 1973; Ceccarelli et al., 1973) and is the subject of several reviews to which the reader is referred (Heuser, 1978; Zimmermann, 1979; Ceccarelli and Hurlbut, 1980; Kelly, 1988). In fully developed synapses, SSVs are clustered at specialised segments of the presynaptic plasma membrane, the active zones, which are in precise register with postsynaptic specializations. In these clusters, they are thought to be locked via synapsin I bridges to a web of actin filaments. When nerve terminals are depolarized, voltage-sensitive calcium channels open, Ca^{++} enters the cell and triggers exocytosis of SSVs. This process is extremely fast (less than 200 µs latency between calcium entry and transmitter release) and tightly controlled by an array of modulatory mechanisms (Augustine et al., 1987; De Camilli and Jahn, 1990). The vesicle membrane is retrieved by endocytosis and used for the reformation of SSVs.

A number of observations indicate that under conditions of moderate activity SSV membranes may not completely flatten into the plasmalemma but rather undergo a very transient fusion sufficient to allow neurotransmitter release (Ceccarelli and Hurlbut, 1980; Valtorta et al., 1988; Torri-Tarelli et al., 1990). They are then retrieved directly at the site of release, without involvement of clathrin-coated vesicles or of an "endosomal" intermediate (Ceccarelli and Hurlbut, 1980; Torri-Tarelli et al., 1987; for a different view see Miller and Heuser, 1984). Under these conditions, no intermixing occurs between SSV membranes and the presynaptic plasma membranes. This is supported by the observation that no SSV membrane proteins are visible in the presynaptic plasma membrane, in spite of electrophysiological evidence for quantal secretion of neurotransmitter (Valtorta et al., 1988).

After intense stimulation, however, the number of SSVs decreases and coated vesicles and cisternae become abundant (Heuser and Reese, 1973). Numerous exocytotic profiles are observed under these conditions when nerve preparations are fixed by ultrarapid freezing (Heuser et al., 1979; Torri-Tarelli et al., 1985). Furthermore, SSV membrane proteins are detectable in the plasma membrane (Von Wedel et al., 1981; Theresa-Jones et al., 1982; Robitaille and Tremblay, 1987; Valtorta et al., 1988; Torri-Tarelli et al., 1990). The intracellular cisternae are labelled by extracellular tracers such as horseradish peroxidase when stimulation is performed in their presence. These cisternae probably represent an "endosomal" compartment of the nerve terminal and eventually disappear in parallel with the reformation of SSVs. This demonstrates that the nerve terminal has the capacity to regenerate SSVs from large precursor membranes (Heuser and Reese, 1973). These cisternae may also represent the organelles responsible for sorting membrane constituents destined to retrograde transport or for accepting SSV proteins delivered by anterograde transport (see section 2.5.1.1).

Regulated exocytosis of LDCVs is less well characterized. Secretory peptides which can only be synthetized in the cell body are degraded after release. Therefore, functional LDCVs cannot be reformed in the nerve terminal. Their membranes must recycle via the Golgi apparatus to be used for another round of exocytosis. At least in the case of the posterior pituitary there is evidence suggesting that the recycling of LDCVs might involve vacuoles in the same size range of LDCVs (Theodosis et al., 1976; Morris and Nordman, 1980). These studies as well as our findings that SSVs and LDCVs have different membrane compositions (Navone et al., 1984; Navone et al., 1986; Baumert et al., 1989; Navone et al., 1989; Baumert et al., 1990), have clearly ruled out that SSVs are directly formed from recycling membranes of LDCVs. This was suggested as a mechanism for SSV formation (Kelly, 1988; Boarder, 1989).

Since the fate of LDCV membranes after exocytosis is largely unknown, it cannot be excluded that endocytosed membrane components of LDCVs are delivered to a sorting compartment common to both SSVs and LDCVs within the nerve terminal (e.g. the large membrane structures mentioned above; see Fig. 2-20a). If this is the case, such a compartment must have the capacity to separate SSV proteins from LDCV proteins and to distinguish between SSV proteins which are designed for re-use in the terminal from those SSV and LDCV membrane proteins which are designed for retrograde transport.

More work is required to unravel the complete membrane cycles of SSVs and LDCVs and the precise interrelationship between them. It cannot be excluded that the concepts outlined above do not apply to specialized systems. For example, the relationship between LDCVs and small dense core vesicles (SSVs of catecholamine containing neurons) present in the varicosities of sympathetic nerves remains to be established. Both types of vesicles are abundant in the varicosities and both contain catecholamines, while only LDCVs contain secretogranin and neuropeptides (Rosa et al., 1985; De Camilli and Jahn, 1990; Thureson-Klein and Klein, 1990). In addition, the membranes of both LDCVs and SSVs in sympathetic varicosities contain the catecholamine carrier, the proton pump and cytochrome b_{561} (Philippu and Matthaei, 1988; Winkler, 1988). A localization of SSV proteins in the varicosities at the electron microscopic level has not yet been reported. In this system, the concept that SSVs are derived from LDCVs still prevails (for a full discussion see Winkler, 1988). Although we favor the view that this system is similar to other neurons, with a clear separation of the pathways for LDCVs and SSVs (Thureson-Klein, 1983; Thureson-Klein and Klein, 1990; De Camilli and Jahn, 1990), more experiments are needed to resolve this issue.

2.5.2.2 Regulation of Exocytosis

There is increasing evidence that exocytosis of SSVs and LDCVs is differentially controlled. The relative proportion of peptide and classical neurotransmitter secretion varies with the pattern of stimulation. It appears that a high frequency of stimulation, interrupted by silent intervals, is optimal for peptide secretion whereas maximal release of classical transmitter is already achieved at lower frequency with continuous stimulation (Lundberg and Hökfelt, 1983; Poulain and Theodosis, 1988). These differences may partially reflect the involvement of different Ca^{++} channels (Rane et al., 1987; Hirning et al., 1988; Smith and Augustine, 1988; Martin and Magistretti, 1989). An almost complete dissociation of LDCV and SSV exocytosis was observed recently in the frog neuromuscular junction. Application of α-latrotoxin led to a complete and irreversible exocytosis of virtually all synaptic vesicles within the nerve terminal. However, the number of LDCVs containing calcitonin-gene related peptide was not significantly decreased (Matteoli et al., 1988; Fig. 2-19). These data suggest that the trigger mechanism for both types of vesicles is different and support the hypothesis that under physiological conditions SSVs and LDCVs respond to the rate and amount of Ca^{++}-increase in a different manner (Smith and Augustine, 1988).

The concept of separate control of SSV and LDCV exocytosis is further supported by the observation that exocytosis of SSVs and LDCVs occurs at different sites. As discussed above, SSVs are concentrated in nerve endings in proximity to active zones. Their release occurs selectively at these sites. In contrast, there is strong evidence indicating that LDCVs undergo exocytosis primarily away from active zones of nerve terminals (Zhu et al., 1986). LDCVs are often found at high concentration in cell bodies and dendrites (Morrison et al., 1984) where they are likely to undergo exocytosis. Within synapses, LDCV exocytosis almost never occurs at the active zone but rather at a more distant location of the plasmalemma (Zhu et al., 1986; Thureson-Klein and Klein, 1990).

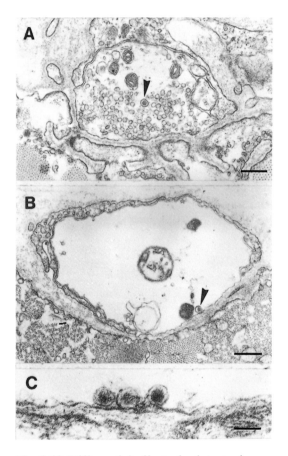

Fig. 2-19 Differential effect of α-latrotoxin on exocytosis of LDCVs and SSVs at the frog neuromuscular junction. (A) shows a control nerve terminal where LDCVs (arrowhead) represent about 1% of the total vesicle content. (B) and (C) show nerve terminals incubated for 1 hr in a Ca^{2+}-free solution in the presence of α-latrotoxin. In these terminals, all SSVs, but not LDCVs, have disappeared as a result of massive exocytosis. A single LDCV (arrowhead) surrounded by an SSV free cytoplasm is visible in (B). Three LDCVs clustered in front of the synaptic cleft are visible in (C). Calibration bars: (A) 340 μm; (B) 529 μm; (C) 178 μm. Reproduced from Matteoli et al. (1988).

2.6 SLMVs and Secretory Granules: Two Pathways of Exocytosis in Endocrine Cells

The discovery of SLMVs (an endomembrane system of endocrine cells and endocrine tumors which shares major membrane proteins with SSVs and which is different from the respective secretory granules) raises the question of how these organelles are related to the classical regulated and constitutive pathway of secretion. As in neurons, the difference in the membrane composition between SLMVs and secretory vesicles excludes a simple precursor-product relationship between the two organelles as suggested by others (Lowe et al., 1988; Kelly, 1988; Boarder, 1989). Instead, we have proposed that SLMVs represent organelles with distinct biogenesis and a completely independent function (Navone et al., 1986; De Camilli and Navone, 1987; De Camilli and Jahn, 1990). This function remains to be determined. An important clue was recently obtained when a fibroblast cell line (CHO cells) which naturally does not express SSV proteins, was transfected with the gene for synaptophysin (Johnston et al., 1989b). In the transfected cells, synaptophysin was incorporated into microvesicles very similar in size and intracellular distribution to SLMVs of endocrine cells. In addition, these vesicles were also labeled by

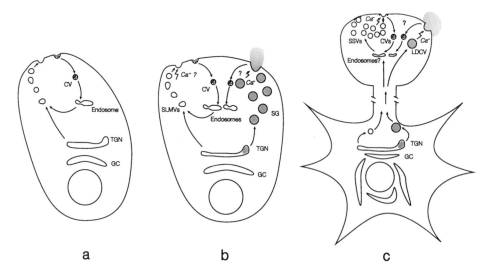

Fig. 2-20 Model of exo-endocytotic pathways in fibroblasts (a), peptide hormone-secreting endocrine cells (b) and neurons (c). SSVs: Small synaptic vesicles; LDCVs: Large dense core vesicles; SLMVs: Synaptic-like microvesicles; TGN: Trans-Golgi Network; GC: Golgi complex; CV: Coated vesicles. See text for details.

extracellular tracers raising the possibility that SLMVs are related to an early endosomal compartment (Johnston et al., 1989b; Fig. 2-18c). Further characterization suggested that these microvesicles are identical with the microvesicles that recycle the transferrin receptor (Johnston et al., 1989b). Since it is likely that a membrane protein expressed in a foreign cell accumulates in the membrane type most closely related to that in which it is normally present, these findings raise the possibility that SLMVs may be related to microvesicles involved in the recycling of plasmalemma receptors. According to this model, SLMVs would represent the endocrine adaptation of a recycling pathway present in all cells and, furthermore, SSVs would represent an even more extreme adaptation of the same pathway (Fig. 2-20). This issue is currently under investigation in our laboratories.

2.7 Conclusions

In this chapter, we have attempted to give an overview of the best characterized membrane proteins of synaptic vesicles and the advancement in the field of neuronal membrane traffic which were possible by the use of these proteins as markers. The study of these proteins has allowed major progress in our understanding of the biogenesis and life cycles of SSVs and, to a lesser extent, of LDCVs, firmly establishing them as functionally and structurally different organelles. In addition, the similarities between SSVs and SLMVs as well as those between LDCVs and secretory granules in neurons and endocrine cells, respectively, has aided the development of new concepts for neuronal secretion in the context of established pathways of exocytosis in other cells. Although the molecular composition of SSVs is presently understood better than that of any other secretory organelle, our picture is far from complete and the characterization of new proteins may quickly outdate this review. Presently, the function of most of the SSV proteins described here is unknown. However, we hope that the detailed knowledge of their structure and their properties will allow functional studies, leading to the elucidation of the molecular mechanisms involved in the exo-endocytotis of these organelles.

Acknowledgements
The authors are greatly indebted to T.C. Südhof (Howard Hughes Medical Institute, Univ. Texas, Dallas, USA), H. Stadler (Max-Planck-Institute for Biophysical Chemistry, Göttingen, FRG), H. Winkler (Department of Pharmacology, Univ. Innsbruck, Austria), M. Treiman (The Panum Institute, University of Copenhagen, Copenhagen, Denmark), N. Brose and P.R.

Maycox (Max-Planck-Institute for Psychiatry, Martinsried, FRG) for helpful discussions and R. H. Scheller (Department of Biological Sciences, Stanford University, Stanford, USA), Å. Thureson-Klein (Department of Pharmacology and Toxicology, University of Mississippi, Jackson, USA) and H. Zimmermann (Institute for Zoology, University of Frankfurt, FRG) for making available manuscript preprints. In addition, we would like to thank our families for their enduring patience for the long hours of absence during the preparation of this chapter. Work in the authors' laboratories was in part supported by grants from the Deutsche Forschungsgemeinschaft to R.J., by grants from Muscular Dystrophy Association and Diabetes Research and Education Foundation to P.D.C.. The support of the Fidia Research Foundation to P.D.C. is also acknowledged.

2.8 References

1 Agoston, D.V., Whittaker, V.P. (1989) Characterization by size, density, osmotic fragiligy, and immunoaffinity, of acetylcholine- and vasoactive intestinal polypeptide-containing storage particles from myenteric neurones of the guinea-pig. *J. Neurochem.* **52,** 1474–1480

2 Al-Awqati, Q. (1986) Proton-translocating ATPases. *Annu. Rev. Cell Biol.* **2,** 179–199

3 Augustine, G.J., Charlton, M.P., Smith, S.J. (1987) Calcium action in synaptic transmitter release. *Annu. Rev. Neurosci.* **10,** 633–693

4 Bader, M. F., Hikita, T., Trifaro, J. M. (1985) Calcium-dependent calmodulin binding to chromaffin granule membranes: Presence of a 65-kilodalton calmodulin-binding protein. *J. Neurochem.* **44,** 526–539

5 Bähler, M., Greengard, P. (1987) Synapsin I bundles F-actin in a phosphorylation-dependent manner. *Nature* **326,** 704–707

6 Bähler, M., Benfenati, F., Valtorta, F., Czernik, A.J., Greengard, P. (1989a) Characterization of synapsin I fragments produced by cysteine-specific cleavage: a study of their interactions with F-actin. *J. Cell Biol.* **108,** 1841–1849

7 Bähler, M., Klein, R.L. and Greengard, P. (1989b) Synapsin Ib is expressed and colocalized with synaptophysin in AtT20 cells. *J. Cell Biol.* **109,** 293a

8 Baines, A.J., Bennett, V. (1985) Synapsin I is a spectrin-binding protein immunologically related to erythrocyte protein 4.1. *Nature* **315,** 410–413

9 Baines, A.J., Bennett, V. (1986) Synapsin I is a microtubule-bundling protein. *Nature* **319,** 145–147

10 Banker, G.A., Waxman, A.B. (1988) Hippocampal neurons generate natural shapes in cell culture. In: *Intrinsic Determinants of Neuronal Form and Function.* (Lasek, R.J., Black, M.M. eds.) Alan R. Liss Inc., New York, pp. 61–82

11 Barbacid, M. (1987) Ras genes. *Annu. Rev. Biochem.* **56,** 779–828

12 Barker, L.A., Dowdall, M.J., Whittaker, V.P. (1973) Choline metabolism in the cerebral cortex of guinea pigs. *Biochem. J.* **130,** 1063–1080

13 Barnekow, A., Jahn, R., Schartl, M.: Synaptophysin: a substrate for the protein tyrosine kinase pp60^c-src in intact synoptic vesicles. *Oncogene,* in press

14 Baumert, M., Maycox, P.R., Navone, F., De Camilli, P., Jahn, R. (1989) Synaptobrevin: An integral membrane protein of 18,000 daltons present in small synaptic vesicles of rat brain. *EMBO J.* **8,** 379–384

15 Baumert, M., Takei, K., Hartinger, H., Burger, P.M., Fischer v. Mollard, G., Maycox, P.R., De Camilli, P., Jahn, R. (1990) P29, a novel tyrosine-phosphorylated membrane protein present in small clear vesicles of neurons and endocrine cells. *J. Cell Biol.* **110,** 1285–1294

16 Benfenati, F., Greengard, P., Brunner, J., Bähler, M. (1989a) Electrostatic and hydrophobic interactions of synapsin I and synapsin I fragments with phospholipid bilayers. *J. Cell Biol.* **108,** 1851–1862

17 Benfenati, F., Bähler, M., Jahn, R., Greengard, P. (1989b) Interactions of synapsin I with small synaptic vesicles: Distinct sites in synapsin I bind to vesicle phospholipids and vesicle proteins. *J. Cell Biol.* **108,** 1863–1872

18 Bishop, A.E., Power, R.F., Polak, J.M. (1988) Markers for neuroendocrine differentiation. *Pathol. Res. Pract.* **183,** 119–128

19 Bixby, J.L., Reichardt, L.F. (1985) The expression and localization of synaptic vesicle antigens at neuromuscular junctions *in vitro. J. Neurosci.* **5,** 3070–3080

20 Boarder, M.R. (1989) Presynaptic aspects of cotransmission: relationship between vesicles and neurotransmitters. *J. Neurochem.* **53,** 1–11

21 Bock, E., Helle, K.B. (1977) Localization of synaptin on synaptic vesicle membranes, synaptosomal plasma membranes and chromaffin granule membranes. *FEBS Lett.* **82,** 175–178

22 Bööj, S., Larsson, P. A., Dahllöf, A. G., Dahlström, A. (1986) Axonal transport of synapsin I- and cholinergic synaptic vesicle-like material; further immunohisto-chemical evidence for transport of axonal cholinergic transmitter vesicles in motor neurons. *Acta Physiol. Scand.* **128,** 155–165

23 Bööj, S., Goldstein, M., Fischer-Colbrie, R., Dahlström, A. (1989) Calcitonin gene-related peptide and chromogranin A: Presence and intra-axonal transport in lumbar motor neurons in the rat, a comparison with synaptic vesicle antigens in immunohistochemical studies. *Neuroscience* **30,** 479–501

24 Bourne, H.R. (1988) Do GTPases direct membrane traffic in secretion? *Cell* **53,** 669–671

25 Bowmann , B.J., Allen, R., Wechser, M.A., Bowman, E.J. (1988a) Isolation of the genes encoding the *Neurospora* vacuolar ATPase: analysis of *vma-2* encoding the 57 kDa polypeptide and comparison to *vma-1. J. Biol. Chem.* **263,** 14002–14007

26 Bowmann E.J., Tenney, K., Bowman, B. (1988b) Isolation of the genes encoding the *Neurospora* vacuolar ATPase: analysis of *vma-1* encoding the 66 kDa subunit reveals homology to other ATPases. *J. Biol. Chem.* **263,** 13994–14001

27 Breer, H., Morris, S.J., Whittaker, V.P. (1977) Adenosine triphosphatase activity associated with purified cholinergic synaptic vesicles of *Torpedo marmorata. Eur. J. Biochem.* **80,** 313–318

28 Brock, T.O., O'Callaghan, J.P. (1987) Quantitative changes in the synaptic vesicle proteins synapsin I and p38 and the astrocyte-specific protein glial fibrillary acidic protein are associated with chemical-induced injury to the rat central nervous system. *J. Neurosci.* **7,** 931–942

29 Brown, D., Hirsch, S., Gluck, S. (1988) An H⁺-ATPase in opposite plasma membrane domains in kidney epithelial cell subpopulations. *Nature* **331,**

30 Browning, M.D., Huang, C.K., Greengard, P. (1987) Similarities between protein IIIa and protein IIIb, two prominent synaptic vesicle-associated phosphoproteins. *J. Neurosci.* **7,** 847–853

31 Buckley, K., Kelly, R.B. (1985) Identification of a transmembrane glycoprotein specific for secretory vesicles of neural and endocrine cells. *J. Cell Biol.* **100,** 1284–1294

32 Buckley, K.M., Floor, E., Kelly, R.B. (1987) Cloning and sequence analysis of cDNA encoding p38, a major synaptic vesicle protein. *J. Cell Biol.* **105,** 2447–2456

33 Buckley, K.M., Kelly, R.B. (1989) Identification and characterization of integral membrane proteins of the synaptic vesicle. *J. Neurochem.* **52,** S21

34 Buffa, R., Rindi, G., Sessa, F., Gini, A., Capella, C., Jahn, R., Navone, F., De Camilli, P., Solcia, E. (1987) Synaptophysin immunoreactivity and small clear vesicles in neuroendocrine cells and related tumours. *Molecular and cellular probes* **1,** 367–381

35 Bunt, A. (1971) Enzymatic digestion of synaptic ribbons in amphibian retinal photoreceptors. *Brain Res.* **25,** 571–577

36 Burger, P.M., Mehl, E., Cameron, P.L., Maycox, P.R., Baumert, M., Lottspeich, F., De Camilli, P., Jahn, R. (1989) Synaptic vesicles immunoisolated from rat cerebral cortex contain high levels of glutamate. *Neuron* **3,** 715–7

37 Burgess, T.L., Kelly, R.B. (1987). Constitutive and regulated secretion of proteins. *Annu. Rev. Cell Biol.,* **3,** 243–293.

38 Burry, R.W., Ho, R.H., Matthew, W.D. (1986) Presynaptic elements formed on polylysine-coated beads contain synaptic vesicle antigens. *J. Neurocytol.* **15,** 409–415

39 Caorsi, C.E., Gonzalez, C.B. (1989) Antibodies against specific membrane proteins of neurosecretory granules isolated from bovine neural lobes. *Neurosci. Lett.* **98,** 241–246

40 Carlson, S.S., Kelly, R.B. (1980) An antiserum specific for cholinergic synaptic vesicles from electric organ. *J. Cell Biol.* **87,** 98–103

41 Carlson, S.S., Kelly, R.B. (1983) A highly antigenic proteoglycan-like component of cholinergic synaptic vesicles. *J. Biol. Chem.* **258,** 11082–11091

42 Carlson, S.S., Caroni, P., Kelly, R.B. (1986) A nerve terminal anchorage protein from electric organ. *J. Cell Biol.* **103,** 509–5

43 Caroni, P., Carlson, S.S., Schweitzer, E., Kelly, R.B. (1985) Presynaptic neurones may contribute a unique glycoprotein to the extracellular matrix at the synapse. *Nature* **314,** 441–443

44 Ceccarelli, B., Hurlbut,W.P. (1980). Vesicle hypothesis of the release of quanta of acetylcholine. *Physiol. Rev.,* **60,** 396–441.

45 Ceccarelli, B., Hurlbut,W.P., Mauro, A. (1973) Turnover of transmitter and synaptic vesicles at the frog neuromuscular junction. *J. Cell Biol.* **57,** 499–524

46 Chardin, P. (1988) The ras superfamily proteins. *Biochimie* **70,** 865–868

47 Chejfec, G., Falkmer, S., Grimelius, L., Jacobsson, B., Rodensjö, M., Wiedenmann, B., Franke,W.W., Lee, I., Gould,V.E. (1987) Synaptophysin. A new marker for pancreatic neuroendocrine tumors. *Am. J. Surg. Pathol.* **11,** 241–247

48 Chun, J.J.M., Shatz, C.J. (1988) Redistribution of synaptic vesicle antigens is correlated with the disappearance of a transient synaptic zone in the developing cerebral cortex. *Neuron* **1,** 297–310

49 Cidon, S., Sihra, T.S. (1989) Characterization of a H$^+$-ATPase in rat brain synaptic vesicles. Coupling to L-glutamate transport. *J. Biol. Chem.* **264,** 8281–8288

50 Cowan, D., Linial, M., Scheller, R.H. (1990) *Torpedo* synaptophysin: Evolution of a synaptic vesicle protein. *Brain Res.* **509,** 1–7

51 Creutz, C.E., Snyder, S.L., Husted, L.D., Beggerly, L.K., Fox, J.W. (1988) Pattern of repeating aromatic residues in synexin. Similarity to the cytoplasmic domain of synaptophysin. Biochem. *Biophys. Res. Commun.* **152,** 1298–1303

52 Czernik, A.J., Pang, D.T., Greengard, P. (1987) Amino acid sequences surrounding the cAMP-dependent and calcium/calmodulin-dependent phosphorylation sites in rat and bovine synapsin I. *Proc. Natl. Acad. Sci. USA* **84,** 7518–7522

53 Dahlström, A.B., Bööj, S. (1988) Rapid axonal transport as a chromatographic process: The use of immunocytochemistry of ligated nerves to investigate the biochemistry of anterogradely versus retrogradely transported organelles. *Cell Motility and Cytoskeleton* **10,** 309–3

54 De Camilli, P., Greengard, P. (1986) Synapsin I: a synaptic vesicle-associated neuronal phosphoprotein. *Biochem. Pharmacol.* **35,** 4349–4357

55 De Camilli, P., Navone, F. (1987) Regulated secretory pathways of neurons and their relation to the regulated secretory pathway of endocrine cells. *Ann. N.Y. Acad. Sci.* **493,** 461–479

56 De Camilli, P., Jahn, R. (1990) Pathways to regulated exocytosis in neurons. *Annu. Rev. Physiol.* **52,** 625–645

57 De Camilli, P., Ueda, T., Bloom, F.E., Battenberg, E., Greengard, P. (1979) Widespread distribution of protein I in the central and peripheral nervous system. *Proc. Natl. Acad. Sci. USA* **76,** 5977–5981

58 De Camilli, P., Cameron, R., Greengard, P. (1983a) Synapsin I (Protein I), a nerve terminal-specific phosphoprotein. I. Its general distribution in synapses of the central and peripheral nervous system demonstrated by immunofluorescence in frozen and plastic sections. *J. Cell Biol.* **96,** 1337–1354

59 De Camilli, P., Harris, S.M., Huttner, W.B., Greengard, P. (1983b) Synapsin I (Protein I), a nerve-terminal-specific phosphoprotein. II. Its specific association with synaptic vesicles demonstrated by immunocytochemistry in agarose-embedded synaptosomes. *J. Cell Biol.* **96,** 1355–1373

60 De Camilli, P., Vitadello, M., Canevini, M.P., Zanoni, R., Jahn, R., Gorio, A. (1988a) The synaptic vesicle proteins synapsin I and synaptophysin (protein p38) are concentrated both in efferent and afferent nerve endings of the skeletal muscle. *J. Neurosci.* **8,** 1625–1631

61 De Camilli, P., Solimena, M., Moretti, M., Navone, F. (1988b) Sites of action of second messengers in the neuronal cytomatrix. In: *Intrinsic determinants of neuronal form and function.* (R.J. Lasek, M.M. Black, eds.) A. R. Liss, Inc., New York, pp. 487–5

62 DeGennaro, L.J., Wallace, W.C., Kanazir, S., Lewis, R.M., Greengard, P. (1983) Neuron-specific phosphoproteins as models for gene expression. *Cold Spring Harbor Symp. Quant. Biol.* **48,** 337–345

63 De Robertis, E., Rodrigues de Lores Arnaiz, G., Salganicoff, L., Pellegrino de Iraldi, A., Zieher, L.M. (1963) Isolation of synaptic vesicles and structural organization of the acetylcholine system within brain nerve endings. *J. Neurochem.* **10,** 225–235

64 Devoto, S.H.; Barnstable, C.J. (1987) SVP38: A synaptic vesicle protein whose appearance correlates closely with synaptogenesis in the rat nervous system. *Ann. N. Y. Acad. Sci.* **493,** 493–496

65 Diebler, M.F., Morot-Gaudry, Y. (1981) Acetylcholine incorporation by cholinergic synaptic vesicles from *Torpedo marmorata. J. Neurochem.* **37**, 467–475

66 Disbrow, J. K., Gershten, M. J., Ruth, J. A. (1982) Uptake of L-[³H] glutamic acid by crude and purified synaptic vesicles from rat brain. *Biochem. Biophys. Res. Commun.* **108**, 1221–1227

67 Eipper, B.A., Mains, R.E. (1988) Peptide α-amidation. *Annu. Rev. Physiol.* **50**, 33–344

68 Eipper, B.A., Park, L.P., Dickerson, I.M., Keutmann, H.T., Thiele, E.A., Rodriguez, H., Schofield, P.R., Mains, R.E. (1987) Structure of the precursor to an enzyme mediating COOH-terminal amidation in peptide biosynthesis. *Mol. Endocrinol.* **1**, 777–790

69 Elferink, L.A., Trimble, W.S., Scheller, R.H. (1989). Two vesicle-associated membrane protein genes are differentially expressed in the rat central nervous system. *J. Biol. Chem.*, **264**, 11061–11064.

70 Esposti, M.D., Kamensky, Y.A., Arutjunjan, A.M., Konstantinov, A.A. (1989) A model for the molecular organization of cytochrome β-561 in chromaffin granule membranes. *FEBS Lett.* **254**, 74–78

71 Favre, D., Scarfone, E., Di Gioia, G., De Camilli, P., Dememes, D. (1986) Presence of synapsin I in afferent and efferent nerve endings of vestibular sensory epithelia. *Brain Res.* **384**, 379–382

72 Fischer v. Mollard, G., Mignery, G., Baumert, M., Perin, M.S., Hanson, T.J., Burger, P.M., Jahn, R., Südhof, T.C. (1990) Rab3 is a small GTP-binding protein exclusively localized to synaptic vesicles. *Proc. Natl. Acad. Sci. USA* **87**, 1988–1992

73 Fisher, J.M., Scheller, R. (1988) Prohormone processing and the secretory pathway. *J. Biol. Chem.* **263**, 16515–16518

74 Fletcher, T.L., De Camilli, P., Banker, G.A. (1989) Synaptogenesis in hippocampal cultures: axons and dendrites become competent to form synapses at different stages of development. *Soc. Neurosci. Abstr.* **15**, 1388

75 Flock A. (1964) Structure of the macula utriculi with special reference to directional interplay of sensory responses as revealed by morphological polarization. *J. Cell Biol.* **22**, 413–431

76 Floor, E., Feist, B.E. (1989) Most synaptic vesicles isolated from rat brain carry three membrane proteins, SV2, synaptophysin, and p65. *J. Neurochem.* **52**, 1433–1437

77 Forgac, M. (1989) Structure and function of vacuolar class of ATP-driven proton pumps. *Physiol. Rev.* **69**, 765–796

78 Forn, J., Greengard, P. (1978) Depolarizing agents and cyclic nucleotides regulate the phosphorylation of specific neuronal proteins in rat cerebral cortex slices. *Proc. Natl. Acad. Sci USA* **73**, 54–58

79 Fournier, S., Trifaro, J.M. (1988) A similar calmodulin-binding protein expressed in chromaffin, synaptic and neurohypophyseal secretory vesicles. *J. Neurochem.* **50**, 27–37

80 Fournier, S., Novas, M.L., Trifaro, J.M. (1989) Subcellular distribution of 65,000 calmodulin-binding protein (p65) and synaptophysin (p38) in adrenal medulla. *J. Neurochem.* **53**, 1043–1049

81 Franke, W.W., Grund, C., Achtstaetter, T. (1986) Co-expression of cytokeratins and neurofilament proteins in a permanent cell line: cultured rat PC 12 cells combine neuronal and epithelial features. *J. Cell Biol.* **103**, 1933–1943

82 Fricker, L.D.(1988) Carboxypeptidase E. *Annu. Rev. Physiol.* **50**, 309–321

83 Fricker, L.D., Evans, C.J., Esch, F.S., Herbert, E. (1986) Cloning and sequence analysis of cDNA for bovine carboxypeptidase E. *Nature* **323**, 461–464

84 Fried, G., Nestler, E.J., De Camilli, P., Stjärne, L., Olson, L., Lundberg, J.M., Hökfelt,T., Ouimet, C.C., Greengard, P. (1982) Cellular and subcellular localization of protein I in the peripheral nervous system. *Proc. Natl. Acad. Sci. USA* **79**, 2717–2721

85 Fuller, R.S., Sterne, R.E., Thorner, J.E. (1988) Enzymes required for yeast prohormone processing. *Annu. Rev. Physiol.* **50**, 345–362

86 Fykse, E., Fonnum, F. (1988) Uptake of γ-aminobutyric acid by a synaptic vesicle fraction isolated from rat brain. *J. Neurochem.* **50**, 1237–1242

87 Gaardsvoll, H., Obendorf, D., Winkler, H., Bock, E. (1988) Demonstration of immunochemical identity between the synaptic vesicle proteins synaptin and synaptophysin/p38. *FEBS Lett.* **242**, 117–1

88 Gasnier, B., Ellory, J.C., Henry, J.P. (1987) Functional molecular mass of binding sites for [^{3}H]dihydrotetrabenazine and [^{3}H]reserpine and of dopamine β-hydroxylase and cytochrome b_{561} from chromaffin granule membrane as determined by radiation inactivation. *Eur. J. Biochem.* **165**, 73–78

89 Giompres, P., Luqmani, Y.A. (1980) Cholinergic synaptic vesicles isolated from *Torpedo marmorata*: Demonstration of acetylcholine and choline uptake in an in vitro system. *Neuroscience* **5**, 1041–1052

90 Goelz, S.E., Nestler, E.J., Cherhazi, B., Greengard, P. (1981) Distribution of Protein I in mammalian brain as determined by a detergent-based radioimmunoassay. *Proc. Natl. Acad. Sci. USA* **78**, 2130–2134

91 Goldenring, J.R., Lasher, R.S., Vollano, M.L., Ueda, T., Naito, S., Sternberger, N.H., Sternberger, L.A., De Lorenzo, R.J. (1986) Association of synapsin I with neuronal cytoskeleton. *J. Biol. Chem.* **261**, 8495–8504

92 Gould,V.E.,Wiedenmann, B., Lee, I., Schwechheimer, K., Dockhorn-Dworniczak, B., Radosevich, J.A., Moll, R., Franke, W.W. (1987) Synaptophysin expression in neuroendocrine neoplasms as determined by immunocytochemistry. *Am. J. Pathol.* **126**, 243–257

93 Gratzl, M.,Torp-Pedersen, C., Dartt, D.,Treiman, M.,Thorn, N.A. (1980) Isolation and characterization of secretory vesicles from bovine neurohypophyses. *Hoppe-Seyler's Z. Physiol. Chem.* **361**, 1615–1628

94 Gray, E.G., Pease, H.L. (1971) On understanding the organization of the retinal receptor synapses. *Brain. Res.* **35**, 1–15

95 Greif, K.F., Reichardt, L.F. (1982) Appearance and distribution of neuronal cell surface and synaptic vesicle antigens in the developing rat superior cervical ganglion. *J. Neurosci.* **2**, 843–852

96 Hancock, J.F.; Magee, A.I.; Childs, J.E.; Marshall, C.J. (1989) All ras proteins are polyisoprenylated but only some are palmitoylated. *Cell* **57**, 1167–1177

97 Harlos, P., Lee, D. A., Stadler, H. (1984) Characterization of a Mg^{2+}-ATPase and a proton pump in cholinergic synaptic vesicles from the electric organ of *Torpedo marmorata*. *Eur. J. Biochem.* **144**, 441–446

98 Havinga, J.R., Strous, G.J., Poort, C. (1984) Intracellular transport of the major glycoprotein of zymogen granule membranes in the rat pancreas. Demonstration of high turnover at the plasma membrane. *Eur. J. Biochem.* **144**, 177–183

99 Havinga, J.R., Slot, J.W., Strous, G.J. (1985) Membrane detachment and release of the major membrane glycoprotein of secretory granules in rat pancreatic exocrine cells. *Eur. J. Cell Biol.* **39**, 70–76

100 Haycock, J.W., Greengard, P., Browning, M.D. (1988) Cholinergic regulation of protein III phosphorylation in bovine adrenal chromaffin cells. *J. Neurosci.* **8,** 3233–3239

101 Hell, J.W., Maycox, P.R., Stadler, H., Jahn, R. (1988) Uptake of GABA by rat brain synaptic vesicles isolated by a new procedure. *EMBO J.* **7,** 3023–3029

102 Heuser, J.E. (1978) Synaptic vesicle exocytosis and recycling during transmitter discharge from the frog neuromuscular junction. In *Transport of Macromolecules in Cellular Systems.* (Silverstein, S.C., ed.) Dahlem Konferenzen, Berlin, pp. 445–464

103 Heuser, J. E., Reese, T. S. (1973) Evidence for recycling of synaptic vesicle membrane during transmitter release at the frog neuromuscular junction. *J. Cell Biol.* **57,** 315–344

104 Heuser, J.E., Reese, T.S., Dennis, M.J., Jan, Y., Jan, L., Evans, L. (1979) Synaptic vesicle exocytosis captured by quick-freezing and correlated with quantal transmitter release. *J. Cell Biol.* **81,** 275–300

105 Hirano, A.A., Greengard, P., Huganir, R.L. (1988). Protein tyrosine kinase activity and its endogenous substrates in rat brain: a subcellular and regional survey. *J. Neurochem.,* **50,** 1447–1455.

106 Hirning, L.D., Fox, A.P., McCleskey, E.W., Olivera, B.M., Thayer, S.A., Miller, R., Tsien, R.W. (1988) Dominant role of N-type Ca^{++} channels in evoked release of norepinephrine from sympathetic neurons. *Science* **239,** 57–61

107 Hirokawa, N., Sobue, K., Kanda, K., Harada, A., Yorifuji, H. (1989) The cytoskeletal architecture of the presynaptic terminal and molecular structure of synapsin I. *J. Cell Biol.* **108,** 111–126

108 Hirsch, S., Strauss, A., Masood, K., Lee, S. Sukhatme,, V., Gluck, S. (1988) Isolation and sequence of a cDNA clone encoding the 31 kDa subunit of bovine kidney vacuolar H^+-ATPase. *Proc. Natl. Acad. Sci. USA* **85,** 3004–3008

109 Hökfelt, T., Fuxe, K., Pernow, P. eds. (1986) Coexistence of neuronal messengers. *Prog. Brain Res.* **68,** 1–411

110 Hooper, J.E., Carlson, S.S., Kelly, R.B. (1980) Antibodies to synaptic vesicles purified from *Narcine* electric organ bind a subclass of mammalian nerve terminals. *J. Cell Biol.* **87,** 104–113

111 Hosie, R.J.A. (1965) The localization of adenosine triphosphatases in morphologically characterized subcellular fractions of guinea-pig brain. *Biochem. J.* **96,** 404–412

112 Howell, T.W., Cockcroft, S., Gomperts, B.D. (1987) Essential synergy between Ca^{2+} and guanine nucleotides in exoytotic secretion from permeabilized rat mast cells. *J. Cell Biol.* **105,** 191–197

113 Huang, C.K., Browning, M.D., Greengard, P. (1982) Purification and characterization of protein IIIb, a mammalian brain phosphoprotein. *J. Biol. Chem.* **257,** 6524–6528

114 Huttner, W.B., DeGennaro, L.J., Greengard, P. (1981) Differential phosphorylation of multiple sites in purified Protein I by cyclic AMP-dependent and calcium-dependent protein kinases. *J. Biol. Chem.* **256,** 1482–1488

115 Huttner, W.B., Schiebler, W., Greengard, P., De Camilli, P. (1983) Synapsin I (Protein I), a nerve terminal-specific phosphoprotein. III. Its association with synaptic vesicles studied in a highly purified synaptic vesicle preparation. *J. Cell Biol.* **96,** 1374–1388

116 Isambert, M.F., Henry, J.P. (1985) Photoaffinity labeling of the tetrabenazine binding sites of bovine chromaffin granule membranes. *Biochemistry* **24,** 3660–3667

117 Isambert, M.F., Gasnier, B., Laduron, P.M., Henry, J.P. (1989) Photoaffinity labeling of the monoamine transporter of bovine chromaffin granules and other monoamine storage vesicles using 7-azido-8-[^{125}I]iodoketanserin. *Biochemistry* **28**, 2265–2270

118 Israel, M., Manaranche, R., Marsal, J., Meunier, F.M., Morel, N., Frachon, P., Lesbats, B. (1980) ATP-dependent calcium uptake by cholinergic synaptic vesicles isolated from *Torpedo* electric organ. *J. Membrane Biol.* **54**, 115–126

119 Jahn, R., Maycox, P. (1988) Protein components and neurotransmitter uptake in brain synaptic vesicles. In: *Molecular Mechanisms in Secretion.* Alfred Benzon Symposium 25 (Thorn, N.A., Treiman, M., Petersen, O.H., eds.) Munksgaard, Copenhagen, pp. 411–424

120 Jahn, R., Schiebler, W., Ouimet, C., Greengard, P. (1985) A 38,000 dalton membrane protein (p38) present in synaptic vesicles. *Proc. Natl. Acad. Sci. USA* **82**, 4137–4141

121 Jahn, R., Navone, F., Greengard, P., De Camilli, P. (1987) Biochemical and immunocytochemical characterization of p38, an integral membrane glycoprotein of small synaptic vesicles. *Ann. N. Y. Acad. Sci.* **493**, 497- 499

122 Jahn, R., Hell, J.W., Maycox, P.R. (1990) Synaptic vesicles: Key organelles involved in neurotransmission. *J. Physiol.* (Paris) **84**, 134–137

123 Janetzko A., Zimmermann, H.,Volknandt,W. (1989) Intraneuronal distribution of a synaptic vesicle membrane protein: Antibody binding sites at axonal membrane compartments and trans-Golgi network and accumulation at nodes of ranvier. *Neuroscience* **32**, 65–77

124 Johnson, E.M., Ueda, T., Maeno, H., Greengard, P. (1972) Adenosine 3',5'-monophosphate-dependent phosphorylation of a specific protein in synaptic membrane fractions from rat cerebrum. *J. Biol. Chem.* **247**, 5650–5652

125 Johnson, R.G. (1988) Accumulation of biological amines into chromaffin granules: A model for hormone and neurotransmitter transport. *Physiol. Rev.* **68**, 232–307

126 Johnston, P.A, Jahn, R., Südhof, T.C. (1989a) Transmembrane topography and evolutionary conservation of synaptophysin. *J. Biol. Chem.* **264**, 1268–1273

127 Johnston, P.A., Cameron, P.L., Stukenbrok, H., Jahn, R., De Camilli, P., Südhof, T.C. (1989b) Synaptophysin is targeted to similar microvesicles in CHO and PC12 cells. *EMBO J.* **8**, 2863–2872

128 Kelly, R.B. (1985) Pathways of protein secretion in eukaryotes. *Science* **230**, 25–32

129 Kelly, R.B. (1988). The cell biology of the nerve terminal. *Neuron,* **1**, 431–438.

130 Kelly, R.B., Deutsch, J.W., Carlson, S.S., Wagner, J.A. (1979) Biochemistry of neurotransmitter release. *Annu. Rev. Neurosci.* **2**, 399–446

131 Kennedy, M.B., Greengard, P. (1981) Two calcium/calmodulin-dependent protein kinases which are highly concentrated in brain, phosphorylate Protein I at distinct sites. *Proc. Natl. Acad. Sci. USA* **72**, 3448–3452

132 Kennedy, M.B., McGuinness, T., Greengard, P. (1983) A calcium/calmodulin-dependent protein kinase activity from mammalian brain that phosphorylates Synapsin I: Partial purification and characterization. *J. Neurosci.* **3**, 818–831

133 Kent, U.M., Fleming, P.J. (1987) Purified cytochrome b$_{561}$ catalyzes transmembrane electron transfer for dopamine β-hydroxylase and peptidyl glycine α-amidating monooxygenase activities in reconstituted systems. *J. Biol. Chem.* **262**, 8174–8178

134 Kidd, M. (1962) Electron microscopy of the inner plexiform layer of the retina in the cat and the pigeon. *J. Anat.* **96**, 179–188

135 Kiene, M.L., Stadler, H. (1987) Synaptic vesicles in electromotoneurones. I. Axonal transport, site of transmitter uptake and processing of a core proteoglycan during maturation. *EMBO J.* **6,** 2209–2215

136 Kish, P.E., Fischer-Bovenkerk, C., Ueda, T. (1989) Active transport of γ-aminobutyric acid and glycine into synaptic vesicles. *Proc. Natl. Acad. Sci. USA* **86,** 3877–3881

137 Klein, R., Lagercrantz, H., Zimmermann, H., eds. (1982) Neurotransmitter Vesicles. Academic Press, New York

138 Knaus, P., Betz, H., Rehm, H. (1986) Expression of synaptophysin during postnatal development of the mouse brain. *J. Neurochem.* **47,** 1302–1304

139 Koenigsberger, R., Parsons, S.M. (1980) Bicarbonate and magnesium ion-ATP dependent stimulation of acetylcholine uptake by *Torpedo* electric organ synaptic vesicles. *Biochem. Biophys. Res. Commun.* **94,** 305–312

140 Kuhn, D.M., Volknandt, W., Stadler, H., Zimmermann, H. (1988) Cholinergic vesicle specific proteoglycan: Stability in isolated vesicles and in synaptosomes during induced transmitter release. *J. Neurochem.* **50,** 11–16

141 Lamouroux, A.,Vigny, A., Faucon Biguet, N., Darmon, M.C., Franck, R., Henry, J.P., Mallet, J. (1987) The primary structure of human dopamine-β-hydroxylase: insights into the relationship between the soluble and the membrane-bound forms of the enzyme. *EMBO J.* **6,** 3931–3937

142 Landis, D.M.D. (1988) Membrane and cytoplasmic structure at synaptic junctions in the mammalian central nervous system. *J. Electron Microscopy Technique* **10,** 129–151

143 Landis, D.M.D., Hall, A.K.,Weinstein, L.A., Reese,T.S. (1988) The organization of cytoplasm at the presynaptic active zone of a central nervous system synapse. *Neuron* **1,** 201–209

144 LeBel, D., Beattie, M. (1988) The major protein of pancreatic zymogen granule membranes (GP-2) is anchored via covalend bonds to phosphatidylinositol. *Biochem. Biophys. Res. Commun.* **154,** 818–823

145 Leclerc, N., Beesley, P.W., Brown, I., Colonnier, M., Gurd, J.W., Paladino, T. and Hawkes, R. (1989) Synaptophysin expression during synaptogenesis in the rat cerebellar cortex. *J. Comp. Neurol.* **280,** 197–212

146 Lee, D.A.,Witzemann,V. (1983) Photoaffinity labeling of a synaptic vesicle specific nucleotide transport system from *Torpedo marmorata*. *Biochemistry* **23,** 6123–6130

147 Lee, D.A., Witzemann, V. (1987) The role of thiols in nucleotide uptake into synaptic vesicles from *Torpedo marmorata*. *Eur. J. Biochem.* **166,** 553–558

148 Leube, R.E., Kaiser, P., Seiter, A., Zimbelmann, R., Franke, W.W., Rehm, H., Knaus, P., Prior, P., Betz, H., Reinke, H., Beyreuther, K.,Wiedenmann, B. (1987) Synaptophysin: molecular organization and mRNA expression as determined from cloned cDNA. *EMBO J.* **6,** 3261–3268

149 Leube, R.E.,Wiedenmann, B., Franke,W.W. (1988) Differentiation markers as an aid in the histological diagnosis of small-cell carcinoma of the lung: synopsis of intermediate filament protein and synaptophysin expression. *Klin.Wochenschr.* **66,** Suppl. XI, 80–86

150 Levitt, P., Rakic, P., De Camilli, P., Greengard, P. (1984) Emergence of cyclic guanosine 3':5'-monophosphate-dependent protein kinase immunoreactivity in developing rhesus monkey cerebellum: correlative immunocytochemical and electron microscopic analysis. *J. Neurosci.* **4,** 2553–2564

151 Lindsley, T.A., De Camilli, P., Banker, G.A. (1987) The influence of cell-cell contact on the distribution of synapsin I in hippocampal neurons in cultures. *Soc. Neurosci. Abstr.* **13,** 318

152 Linial, M., Miller, K., Scheller, R.H. (1989) VAT-1: An abundant membrane protein from *Torpedo* cholinergic synaptic vesicles. *Neuron* **2,** 1265–1273

153 Llinas, R., McGuinness, T.L., Leonard, C.S., Sugimori, M., Greengard, P. (1985) Intraterminal injection of synapsin I or calcium/calmodulin-dependent protein kinase II alters neurotransmitter release at the squid giant synapse. *Proc. Natl. Acad. Sci. USA* **82,** 3035–3039

154 Loh, Y.P. (1987) Processing of pro-opiomelanocortin and other peptide precursors by unique, secretory vesicle enzymes. In: *Molecular Mechanisms in Secretion.* Alfred Benzon Symposium 25 (Thorn, N.A., Treiman, M., Petersen, O.H., eds) Munksgaard, Copenhagen, pp. 525–539

155 Lohmann, S.M., Ueda, T., Greengard, P. (1978) Ontogeny of synaptic phosphoproteins in brain. *Proc. Natl. Acad. Sci. USA* **75,** 4037–4041

156 Lowe, A.W., Madeddu, L., Kelly, R.B. (1988) Endocrine secretory granules and neuronal synaptic vesicles have three integral membrane proteins in common. *J. Cell Biol.* **106,** 51–59

157 Lundberg, J.M., Hökfelt, T. (1983) Coexistence of peptides and classical neurotransmitters. *Trends Neurosci.* **6,** 325–333

158 Luqmani, Y.A. (1981) Nucleotide uptake by isolated cholinergic synaptic vesicles: Evidence for a carrier of adenosine 5'-triphosphate. *Neuroscience* **6,** 1011–1021

159 Mains, R.E., Dickerson, I.M., May, V., Stoffers, D.A., Perkins, S.N., Ouafik, L., Husten, E.J., Eipper, B.A. (1990) Cellular and molecular aspects of peptide hormone biosynthesis. *Frontiers in Neuroendocrinology,* in press

160 Mandel, M., Moriyama, Y., Hulmes, J.D., Pan, Y.E., Nelson, H., Nelson, N. (1988) cDNA sequence encoding the 16-kDa proteolipid of chromaffin granules implies gene duplication in the evolution of H^+-ATPases. *Proc. Natl. Acad. Sci. USA* **85,** 5521–5524

161 Mandell, J.W., Townes-Anderson, E., Cameron, R., Greengard, P. De Camilli, P. (1989) Retinal ribbon synapses lack synapsin immunoreactivity. *Soc. Neurosci. Abstr.* **15,** 680

162 Manolson, M.F., Ouellette, B.F.F., Filion, M., Poole, R.J. (1988) cDNA sequence and homologies of the "57-kDa" nucleotide-binding subunit of the vacuolar ATPase from *Arabidopsis. J. Biol. Chem.* **263,** 17987–17994

163 March, P.E., Antonian, E., Schneider, D., Rothwarf, D.M., Thornton, E.R. (1987) The high-molecular weight proteins of bovine brain plain synaptic vesicles. *Neurochem. Res.* **12,** 635–640

164 Marshall, I.G., Parsons, S.M. (1987) The vesicular acetylcholine transport system. *Trends Neurosci.* **10,** 174–177

165 Martin, J.L., Magistretti, P.J. (1989) Pharmacological studies of the voltage-sensitive Ca^{2+} channels involved in the release of vasoactive intestinal peptide evoked by K^+ in mouse cerebral cortical slices. *Neuroscience* **30,** 423–431

166 Mason, C.A. (1986) Axon development in mouse cerebellum: embryonic axon forms and expression of synpasin I. *Neuroscience* **19,** 1319–1333

167 Matsui, Y., Kikuchi, A., Kondo, J., Hishida, T., Teranishi, T., Takai, Y. (1988) Nucleotide and deduced amino acid sequences of a GTP-binding protein family with molecular weights of 25,000 from bovine brain. *J. Biol. Chem.* **263,** 11071–11074

168 Matteoli, M., Haimann, C., Torri-Tarelli, F., Polak, J.M., Ceccarelli, B., De Camilli, P. (1988) Differential effect of α-latrotoxin on exocytosis from small synaptic vesicles and from large dense-core vesicles containing calcitonin gene-related peptide at the frog neuromuscular junction. *Proc. Natl. Acad. Sci. USA* **85**, 7366–7370

169 Matthew, W.D., Tsavaler, L., Reichardt, L.F. (1981) Identification of a synaptic vesicle-specific membrane protein with a wide distribution in neuronal and neurosecretory tissue. *J. Cell Biol.* **91**, 257–269

170 Maycox, P. R., Deckwerth, T., Hell, J.W., Jahn, R. (1988) Glutamate uptake by brain synaptic vesicles: Energy dependence of transport and functional reconstitution in proteoliposomes. *J. Biol. Chem.* **263**, 15423–15428

171 Maycox, P.R., Hell, J.W., Jahn, R. (1990) Amino acid neurotransmission: Spotlight on synaptic vesicles. *Trends Neurosci.* **13**, 83–87

172 McCaffery, C. A., DeGennaro, L. J. (1986) Determination and analysis of the primary structure of the nerve terminal specific phosphoprotein, synapsin I. *EMBO J.* **5**, 3167–3173

173 Mellman, I., Fuchs, R., Helenius, A. (1986) Acidification of the endocytic and exocytic pathways. *Annu. Rev. Biochem.* **55**, 663–700

174 Mellman, I., Howe, C., Helenius, A. (1987) The control of membrane traffic on the endocytic pathway. *Current Topics in Membranes and Transport* **29**, 255–289

175 Metz, J., Gerstheimer, F.P., Herbst, M. (1986) Distribution of synaptophysin immunoreactivity in guinea pig heart. *Histochemistry* **86**, 221–224

176 Michaelson, D.M., Ophir, I., Angel, I. (1980) ATP-stimulated Ca^{2+} transport into cholinergic *Torpedo* synaptic vesicles. *J. Neurochem.* **35**, 116–124

177 Miettinen, M. (1987) Synaptophysin and neurofilament proteins as markers for neuroendocrine tumors. *Arch. Pathol. Lab. Med.* **111**, 813–818

178 Miller, T.M., Heuser, J.E. (1984) Endocytosis of synaptic vesicle membrane at the frog neuromuscular junction. *J. Cell Biol.* **98**, 685–698

179 Mizoguchi, A.; Kim, S.; Ueda, T.; Takai, Y. (1989) Tissue distribution of smg p25A, a ras p21-like GTP-binding proteins, studied by use of a specific monoclonal antibody. Biochem. *Biophys. Res. Commun.* **162**, 1438–1445

180 Morris, J.F., Nordman, J.J. (1980) Membrane recapture after hormone release from nerve endings in the neural lobe of the rat pituitary gland. *Neuroscience* **5**, 639–649

181 Morris, J.F., Chapman, D.B., Sokol, H.W. (1987) Anatomy and function of classic vasopressin-secreting hypothalamus-neurohypophysial system. In: *Vasopressin. Principles and properties.* (Gash, D.M., Boer, G.J., eds.) Plenum, New York, pp. 1–89

182 Morrison, J.H., Magistretti, P.J., Benoit, R., Bloom, F.E. (1984) The distribution and morphological characteristics of the intracortical VIP-positive cells: an immunohistochemical analysis. *Brain Res.* **292**, 269–282

183 Nagy, A., Baker, R.R., Morris, S.J., Whittaker, V.P. (1976) The preparation and characterization of synaptic vesicles of high purity. *Brain Res.* **109**, 285–309

184 Nairn, A.C., Greengard, P. (1987) Purification and characterization of Ca^{2+}/calmodulin-dependent protein kinase I from bovine brain. *J. Biol. Chem.* **262**, 7273–7281

185 Naito, S., Ueda, T. (1983) Adenosine triphosphate-dependent uptake of glutamate into protein I-associated synaptic vesicles. *J. Biol. Chem.* **258**, 696–699

186 Navone, F., Greengard, P., De Camilli, P. (1984) Synapsin I in nerve terminals: Selective association with small synaptic vesicles. *Science* **226**, 1209–1211

187 Navone, F., Jahn, R., Di Gioia, G., Stukenbrok, H., Greengard, P., De Camilli, P. (1986) Protein p38: An integral membrane protein specific for small vesicles of neurons and neuroendocrine cells. *J. Cell Biol.* **103**, 2511–2527

188 Navone, F., DiGioia, G., Matteoli, M., De Camilli, P. (1988) Small synaptic vesicles and large dense-core vesicles of neurons are related to two distinct types of vesicles of endocrine cells. In: *Molecular mechanisms in secretion. Alfred Benzon Symposium 25* (Thorn, N.A., Treiman, M., Petersen, O.H. eds.). Munksgaard, Copenhagen, pp.433–450

189 Navone, F., Di Gioia, G., Jahn, R., Browning, M., Greengard, P., De Camilli, P. (1989) Microvesicles of the neurohypophysis are biochemically related to small synaptic vesicles of presynaptic nerve terminals. *J. Cell Biol.,* **109**, 3425–3433

190 Nelson, N., Taiz, L. (1989) The evolution of H^+-ATPases. *Trends Biochem.* **14**, 113–116

191 Nelson, H., Mandiyan, S., Nelson, N. (1989) A conserved gene encoding the 57-kDa subunit of the yeast vacuolar H^+-ATPase. *J. Biol. Chem.* **264**, 1775–1778

192 Nestler, E.J., Greengard, P. (1984) *Protein phosphorylation in the nervous system.* John Wiley & Sons, New York, pp. 1–398

193 Njus, D., Kelley, P.M., Harnadek, G.J. (1986) Bioenergetics of secretory vesicle transport. *Biochim. Biophys. Acta* **853**, 237–256

194 Nordmann, J.J., Louis, F., Legros, J.J. (1979) Purification of two structurally and morphologically distinct populations of rat neurohypophysial neurosecretory granules. *Neuroscience* **4**, 1367–1379

195 Obata, K., Nishiye, H., Fujita, S., Shirao, T., Inoue, H., Uchizono, K. (1986) Identification of a synaptic vesicle-specific 38,000 dalton protein by monoclonal antibodies. *Brain Res.* **375**, 37–48

196 Obata, K., Kojima, N., Nishiye, H., Inoue, H., Shirao, T., Fujita, S.C., Uchizono, K. (1987) Four synaptic vesicle-specific proteins: identification by monoclonal antibodies and distribution in the nervous tissue and the adrenal medulla. *Brain Res.* **404**, 169–179

197 Obendorf, D., Schwarzenbrunner, U., Fischer-Colbrie, R., Laslop, A., Winkler, H. (1988a) Immunological characterization of a membrane glycoprotein of chromaffin granules: its presence in endocrine and exocrine tissues. *Neuroscience* **25**, 343–351

198 Obendorf, D., Schwarzenbrunner, U., Fischer-Colbrie, R., Laslop, A., Winkler, H. (1988b) In adrenal medulla synaptophysin (protein p38) is present in chromaffin granules and in a special vesicle population. *J. Neurochem.* **51**, 1573–1580

199 Obendorf, D., Schwarzenbrunner, U., Fischer-Colbrie, R., Scherman, D., Hook, V., Winkler, H. (1989) Do large and small dense core vesicles of sympathetic nerve contain the same antigens? *J. Neurochem.* **52, Suppl.,** 84

200 Olofsson, B., Chardin, P., Touchot, N., Zahraoui, A., Tavitian, A. (1988) Expression of the ras-related ralA, rho12 and rab genes in adult mouse tissues. *Oncogene* **3**, 231–234

201 Ovchinnikov, Y.A., Abdulaev, N.G., Zolotarev, A.S., Artamonov, I.D., Bespalov, I.A., Dergachev, A.E., Tsuda, M. (1988) *Octopus* rhodopsin: Amino acid sequence deduced from cDNA. *FEBS Lett.* **232**, 69–72

202 Pang, D.T., Wang, J.K.T., Valtorta, F., Benfenati, F., Greengard, P. (1988a) Protein tyrosine phosphorylation in synaptic vesicles. *Proc. Natl. Acad. Sci. USA* **85**, 762–766

203 Pang, D.T.,Valtorta, F., DeMarco, M.E., Brugge, J.S., Benfenati, F., Greengard, P. (1988b) Localization of pp60[c-src] in synaptic vesicles. *Soc. Neurosci. Abstr.* **14,** 106

204 Paquette, J., Leblond, F.A., Beattie, M., LeBel, D. (1986) Reducing conditions induce a total degradation of the major zymogen granule membrane protein in both its membranous and its soluble form. Immunochemical quantitation of the two forms. *Biochem. Cell Biol.* **64,** 456–462

205 Pearson, R.B.,Woodgett, J.R., Cohen, P., Kemp, B.E. (1985) Substrate specificity of a multifunctional calmodulin-dependent protein kinase. *J. Biol. Chem.* **260,** 14471–14476

206 Perin, M.S., Fried, V. A., Mignery, G. A., Jahn, R., Südhof, T. C. (1990) Phospholipid binding by a synaptic vesicle protein homologous to the regulatory domain of protein kinase C. *Narue* **345,** 260–263

207 Perin, M.S., Fried, V.A., Slaughter, C.A., Südhof, T.C. (1988) The structure of cytochrome b561, a secretory vesicle-specific electron transport protein. *EMBO J.* **7,** 2697–2703

208 Peters, A., Palay, S.L., Webster, H.F. (1976) *The fine structure of the nervous system. The neurons and supporting cells.* W.B. Saunders, Philadelphia.

209 Petrucci, T.C., Morrow, J.S. (1987) Synapsin I: An actin-bundling protein under phosphorylation control. *J. Cell Biol.* **105,** 1355–1363

210 Philippu, A., Matthaei, H. (1988) Transport and storage of catecholamines in vesicles. *Handbook of Experimental Pharmacology* **90,** 1–42

211 Popoli, M., Mengano, A. (1988) A hemagglutinin specific for sialic acids in a rat brain synaptic vesicle-enriched fraction. *Neurochem. Res.* **13,** 63–67

212 Popoli, M., Mengano, A., Campanella, G. (1989) Ganglioside binding proteins of synaptic vesicles. Are they involved in synaptic transmission? *J. Neurochem.* **52,** S35

213 Poulain, D.A.,Theodosis, D.T. (1988) Coupling of electrical activity and hormone release in mammalian neurosecretory neurons. *Current Topics in Neuroendocrinology* **9,** 73–104

214 Pruss, R.M. (1987) Monoclonal antibodies to chromaffin cells can distinguish proteins specific to or specifically excluded from chromaffin granules. *Neuroscience* **22,** 141–147

215 Pruss, R.M., Shepard, E.A. (1987) Cytochrome b$_{561}$ can be detected in many neuroendocrine tissues using a specific monoclonal antibody. *Neuroscience* **22,** 149,-157

216 Rahamimoff, R., DeRiemer, S.A., Sakmann, B., Stadler, H.,Yakir, N. (1988) Ion channels in synaptic vesicles from *Torpedo* electric organ. *Proc. Natl. Acad. Sci. USA* **85,** 5310–5314

217 Rane, S, Holz, G.G., Dunlap, K. (1987) Dihydropyridine inhibition of neuronal calcium current and substance P release. *Pflüg. Arch.* **409,** 361–366

218 Rehm, H., Wiedenmann, B., Betz, H. (1986) Molecular characterization of synaptophysin, a major calcium-binding protein of the synaptic vesicle membrane. *EMBO J.* **5,** 535–541

219 Reichardt, L.F., Kelly, R.B. (1983). A molecular description of nerve terminal function. *Annu. Rev. Biochem.,* **52,** 871–926

220 Roberts, A., Bush, B.M.H. (eds.)(1981) *Neurons without impulses.* Cambridge Univ. Press, Cambridge, pp. 1–290

221 Robitaille, B., Tremblay, J.P. (1987) Incorporation of vesicular antigens into the presynaptic membrane during exocytosis at the frog neuromuscular junction: A light and electron microscopy immunochemical study. *Neuroscience* **21,** 619–629

222 Romano, C., Nichols, R.A., Greengard, P., Greene, L.A. (1987) Synapsin I in PC12 cells. I. Characterization of the phosphoprotein and effect of chronic NGF treatment. *J. Neurosci.* **7**, 1294–1299

223 Rosa, P., Hille, A., Lee, R.W.H., Zanini, A., De Camilli, P., Huttner, W.B. (1985) Secretogranins I and II: Two tyrosine-sulfated secretory proteins common to a variety of cells secreting peptides by the regulated pathway. *J. Cell Biol.* **101**, 1999–2011

224 Rudnick, G. (1986) ATP-driven H^+ pumping into intracellular organelles. *Annu. Rev. Physiol.* **48**, 403–413

225 Russell, J.T. (1980) The isolation of purified neurosecretory vesicles from bovine neurohypophysis using isoosmolar density gradients. *Anal. Biochem.* **113**, 229–238

226 Russell, J.T. (1987) The secretory vesicle in processing and secretion of neuropeptides. *Current Topics in Membranes and Transport* **31**, 277–312

227 Salminen, A., Novick, P.J. (1987) A ras-like protein is required for a post-Golgi event in yeast secretion. *Cell* **49**, 527–538

228 Sano, K.; Kikuchi, A.; Matsui, Y.; Ternaishi, Y.; Takai, Y. (1989) Tissue-specific expression of a novel GTP-binding protein (smg p25A) mRNA and its increase by nerve growth factor and cyclic AMP in rat pheochromocytoma PC-12 cells. *Biochem. Biophys. Res. Commun.* **158**, 377–385

229 Scarfone, E., Demèmes, D., Jahn, R., De Camilli, P., Sans, A. (1988) Secretory function of the vestibular nerve calyx suggested by presence of vesicles, synapsin I and synaptophysin. *J. Neurosci.* **8**, 4640–4645

230 Scharschmidt, B.F., Van Dyke, R.W. (1987) Proton transport by hepatocyte organelles and isolated membrane vesicles. *Annu. Rev. Physiol.* **49**, 69–85

231 Schiebler, W., Jahn, R., Doucet, J.P., Rothlein, J., Greengard, P. (1986) Characterization of synapsin I binding to small synaptic vesicles. *J. Biol. Chem.* **261**, 8383–8390

232 Schilling, K., Gratzl, M. (1988) Quantification of p38/synaptophysin in highly purified adrenal medullary chromaffin vesicles. *FEBS Lett.* **233**, 22–24

233 Schmid, A., Burckhardt, G., Gögelein, H. (1989) Single chloride channels in endosomal vesicle preparations from rat kidney cortex. *J. Membrane Biol.* **111**, 265–275

234 Schmitt, H.D.; Puzicha, M.; Gallwitz, D. (1988) Study of a temperature-sensitive mutant of the ras-related YPT1 gene product in yeast suggests a role in the regulation of intracellular calcium. *Cell* **53**, 635–647

235 Schneider, D.L. (1987) The proton pump ATPase of lysosomes and related organelles of the vacuolar apparatus. *Biochim. Biophys. Acta* **895**, 1–10

236 Segev, N.; Mulholland, J., Botstein, D. (1988) The yeast GTP-binding YPT1 protein and a mammalian counterpart are associated with the secretion machinery. *Cell* **52**, 915–924

237 Sihra, T.S.; Wang, J.K.T.; Gorelick, F.S.; Greengard, P. (1989) Translocation of synapsin I in response to depolarization of isolated nerve terminals. *Proc. Natl. Acad. Sci. USA* **86**, 8018–8112

238 Sjöstrand, F.S. (1953) Ultrastructure of the outer segments of rods and cones of the eye as revealed by the electron microscope. *Cell. Comp. Physiol.* **42**, 15–44

239 Smith, A.D., Winkler, H. (1967) A simple method for the isolation of adrenal chromaffin granules on a large scale. *Biochem. J.* **103**, 480–482

240 Smith, A.P., Loh, H.H. (1981) Organization of brain synaptic vesicle proteins. *J. Neurochem.* **36**, 1749–1757

241 Smith, C.A., Sjöstrand, F.S. (1961) A synaptic structure in the hair cells of the guinea pig cochlea. *J. Ultrastruct. Res.* **5**, 523–556

242 Smith, S.J., Augustine, G.J. (1988). Calcium ions, active zones and synaptic transmitter release. *Trends Neurosci.* **11**, 458–464.

243 Sossin, W.S., Fisher, J.M., Scheller, R.H. (1989) Cellular and molecular biology of neuropeptide processing and packaging. *Neuron* **2**, 1407–1417

244 Stadler, H., Dowe, G.H.C. (1982) Identification of a heparan sulfate-containing proteoglycan as a specific core component of cholinergic synaptic vesicles from *Torpedo marmorata*. *EMBO J.* **1**, 1381–1384

245 Stadler, H., Fenwick, E.M. (1983) Cholinergic synaptic vesicles from *Torpedo marmorata* contain an atractyloside-binding protein related to the mitochondrial ADP/ATP carrier. *Eur. J. Biochem.* **136**, 377–382

246 Stadler, H., Kiene, M.L. (1987) Synaptic vesicles in electromotoneurones. II. Heterogeneity of populations is expressed in uptake properties; exocytosis and insertion of a core proteoglycan into the extracellular matrix. *EMBO J.* **6**, 2217–2221

247 Stadler, H., Tashiro, T. (1979) Isolation of synaptosomal plasma membranes from cholinergic nerve terminals and a comparison of their proteins with those of synaptic vesicles. *Eur. J. Biochem.* **101**, 171–178

248 Stadler, H., Tsukita, S. (1984) Synaptic vesicles contain an ATP-dependent proton pump and show 'knob-like' protrusions on their surface. *EMBO J.* **3**, 3333–3337

249 Steiner, J.P., Ling, E., Bennett, V. (1987) Nearest neighbor analysis for brain synapsin I. Evidence from in vitro reassociation assays for association with membrane protein (s) and the $M_r = 68,000$ neurofilament subunit. *J. Biol. Chem.* **262**, 905–914

250 Stoffers, D.A., Green, C.B.A., Eipper, B.A. (1989) Alternative mRNA splicing generates multiple forms of peptidyl-glycine α-amidating monooxygenase in rat atrium. *Proc. Natl. Acad. Sci. USA* **86**, 735–739

251 Südhof, T.C., Lottspeich, F., Greengard, P., Mehl, E., Jahn, R. (1987) A synaptic vesicle protein with a novel cytoplasmic domain and four transmembrane regions. *Science* **238**, 1142–1144

252 Südhof, T.C., Czernik, A.J., Kao, H.T., Takei, K., Johnston, P.A., Horiuchi, A., Kanazir, S.D., Wagner, M.A., Perin, M.S., De Camilli, P., Greengard, P. (1989a) Synapsins: Mosaics of shared and individual domains in a family of synaptic vesicle phosphoproteins. *Science* **245**, 1474–1480

253 Südhof, T.C., Baumert, M., Perin, M.S., Jahn, R. (1989b) A synaptic vesicle membrane protein is conserved from mammals to *Drosophila*. *Neuron* **2**, 1475–1481

254 Südhof, T.C., Fried, V.A., Stone, D.K., Johnston, P.A., Xie, X.S. (1989c) Human endomembrane H$^+$ pump strongly resembles the ATP-synthetase of archaebacteria. *Proc. Natl. Acad. Sci. USA* **86**, 6067–6071

255 Sun, S.Z., Xie, X.S., Stone, D.K. (1987) Isolation and reconstitution of the dicyclohexylcarbodiimide-sensitive proton pore of the clathrin-coated vesicle proton translocating complex. *J. Biol. Chem.* **262**, 14790–14794

256 Taylor, C.S., Kent, U.M. and Fleming, P.J. (1989) The membrane-binding segment of dopamine β-hydroxylase is not an uncleaved signal sequence. *J. Biol. Chem.* **264**, 14–16

257 Theodosis, D.T., Dreifuss, J.J., Harris, M.C., Orci, L. (1976) Secretion-related uptake of horseradish peroxidase in neurohypophysial axons. *J. Cell Biol.* **70**, 294–303

258 Theresa-Jones, R., Walker, J.H., Stadler, H., Whittaker, V.P. (1982) Immunohistochemical localization of a synaptic-vesicle antigen in a cholinergic neuron under conditions of stimulation and rest. *Cell Tiss. Res.* **223,** 117–126

259 Thomas, L., Hartung, K., Langosch, D., Rehm, H., Bamberg, E., Franke, W.W., Betz, H. (1988a) Identification of synaptophysin as a hexameric channel protein of the synaptic vesicle membrane. *Science* **242,** 1050–1053

260 Thomas, G., Thorne, B.A., Thomas, L., Allen, R.G., Hruby, D.E., Fuller, R., Thorner, J. (1988b) Yeast KEX2 endopeptidase correctly cleaves a neuroendocrine prohormone in mammalian cells. *Science* **241,** 226–230

261 Thureson-Klein, Å. K. (1983) Exocytosis from large and small dense cored vesicles in noradrenergic nerve terminals. *Neuroscience* **10,** 245–252

262 Thureson-Klein, Å.K., Klein, R.L. (1990) Exocytosis from neuronal large dense-cored vesicles. *Int. Rev. Cytol.* **121,** 67–126

263 Tixier-Vidal, A., Faivre-Bauman, A., Picart, R., Wiedenmann, B. (1988) Immunoelectron microscopic localization of synaptophysin in a Golgi subcompartment of developing hypothalamic neurons. *Neuroscience* **26,** 847–861

264 Tomlinson, D.R. (1975) Two populations of granular vesicles in constricted post-ganglionic sympathetic nerves. *J. Physiol.* (Lond.) **245,** 727–735

265 Tooze, J., Hollinshead, M., Fuller, S.D., Tooze, S.A., Huttner, W.B. (1989) Morphological and biochemical evidence showing neuronal properties in AtT-20 cells and their growth cones. *Eur. J. Cell Biol.* **49,** 259–273

266 Torri-Tarelli, F., Grohovaz, F., Fesce, R., Ceccarelli, B. (1985) Temporal coincidence between synaptic vesicle fusion and quantal secretion of acetylcholine. *J. Cell Biol.* **101,** 1386–1399

267 Torri-Tarelli, F., Haimann, C., Ceccarelli, B. (1987) Coated vesicles and pits during enhanced quantal release of acetylcholine at the neuromuscular junction. *J. Neurocytol.* **16,** 205–214

268 Torri-Tarelli, F., Villa, A., Valtorta, F., De Camilli, P., Greengard, P., Ceccarelli, B. (1990) Redistribution of synaptophysin and synapsin I during α-latrotoxin-induced release of neurotransmitter at the neuromuscular junction. *J. Cell Biol.* **110,** 449–459

269 Touchot, N.; Chardin, P.; Tavitian, A. (1987) Four additional members of the ras gene superfamily isolated by an oligonucleotide strategy: molecular cloning of YPT-related cDNAs from a rat brain library. *Proc. Natl. Acad. Sci. USA* **84,** 8210–8214

270 Trifaro, J.-M., Fournier, S., Novas, M.L. (1989) The p65 protein is a calmodulin-binding protein present in several types of secretory vesicles. *Neuroscience* **29,** 1–8

271 Trimble, W.S., Cowan, D.M., Scheller, R.H. (1988) VAMP-1: A synaptic vesicle-associated integral membrane protein. *Proc. Natl. Acad. Sci. USA* **85,** 4538–4542

272 Tsukita S., Ishikawa, H. (1980) The movement of membraneous organelles in axons. Electron microscopic identification of anterogradely and retrogradely transported organelles. *J. Cell Biol.* **84,** 513–530

273 Ueda, T., Greengard, P. (1977) Adenosine 3':5'-monophosphate-regulated phosphoprotein system of neuronal membranes. I. Solubilization, purification, and some properties of an endogenous phosphoprotein. *J. Biol. Chem.* **252,** 5155–5163

274 Ueda, T., Maeno, H., Greengard, P. (1973) Regulation of endogenous phosphorylation of specific proteins in synaptic membrane fractions from rat brain by adenosine 3':5'-monophosphate. *J. Biol. Chem.* **248,** 8295–8305

275 Vallar, L., Biden, T.J., Wollheim, C.B. (1987) Guanine nucleotides induce Ca^{2+}-independent insulin secretion from permeabilized RINM5F cells. *J. Biol. Chem.* **262**, 5049–5056

276 Valtorta, F., Jahn, R., Fesce, R., Greengard, P., Ceccarelli, B. (1988) Synaptophysin (p38) at the frog neuromuscular junction: its incorporation into the axolemma and recycling after intense quantal secretion. *J. Cell Biol.* **107**, 2717–2727

277 Vetter, J., Betz, H. (1989) Expression of synaptophysin in the rat pheochromocytoma cell-line PC12. Exp. Cell Res. 184, 360–366

278 Volknandt, W., Naito, S., Ueda, T., Zimmermann, H. (1987) Synapsin I is associated with cholinergic nerve terminals in the electric organs of *Torpedo*, *Electrophorus and Malapterurus* and copurifies with *Torpedo* synaptic vesicles. *J. Neurochem.* **49**, 342–347

279 Volknandt, W., Henkel, A., Zimmermann, H. (1988) Heterogeneous distribution of synaptophysin and protein 65 in synaptic vesicles isolated from rat cerebral cortex. *Neurochem. Int.* **12**, 337–345

280 Volknandt, W., Schläfer, M., Zimmermann, H. (1989) SVP25: An integral membrane glycoprotein of synaptic vesicles. *Biol. Chem. Hoppe-Seyler* **370**, 1005

281 Von Schwarzenfeld, I. (1979) Origin of transmitter released by electrical stimulation from a small, metabolically very active vesicular pool of the cholinergic synapses in guinea-pig cerebral cortex. *Neuroscience* **4**, 477–493

282 Von Wedel, R.J., Carlson, S.S., Kelly, R.B. (1981) Transfer of synaptic vesicle antigens to the presynaptic plasma membrane during exocytosis. *Proc. Natl. Acad. Sci. USA* **78**, 1014–1018

283 Walaas, S.I., Jahn, R., Greengard, P.(1988) Quantitation of nerve terminal populations: Synaptic vesicle-associated proteins as markers for synaptic density in the rat neostriatum. *Synapse* **2**, 516–5

284 Walker, J.H., Obrocki, J., Zimmermann, C.W. (1983) Identification of a proteoglycan antigen characteristic of cholinergic synaptic vesicles. *J. Neurochem.* **41**, 209–216

285 Walker, J.H., Kristjansson, I., Stadler, H. (1986) Identification of a synaptic vesicle antigen (M_r 86,000) conserved between *Torpedo* and rat. *J. Neurochem.* **46**, 875–881

286 Walworth, N.C., Goud, B., Kabcenell, A.K., Novick, P.J. (1989) Mutational analysis of SEC4 suggests a cyclical mechanism for the regulation of vesicular traffic. *EMBO J.* **8**, 1685–1693

287 Wang, S.Y., Moriyama, Y., Mandel, M., Hulmes, J.D., Pan, Y.C.E., Danho, W., Nelson, H., Nelson, N. (1988) Cloning of cDNA encoding a 32-kDa protein. An accessory polypeptide of the H^+-ATPase from chromaffin granules. *J. Biol. Chem.* **263**, 17638–17642

288 Weiss, S., Pin, J.P., Sebben, M., Kemp, D., Sladeczek, F., Gabrion, J., Bockaert, J. (1986) Synaptogenesis of cultured striatal neurons in serum-free medium: A morphological and biochemical study. *Proc. Natl. Acad. Sci. USA* **83**, 2238–2242

289 Wharton, J., Gulbenkian, S., Merighi, A., Kuhn, D.M., Jahn, R., Taylor, K.M., Polak, J.M. (1988) Immunohistochemical and ultrastructural localisation of peptide-containing nerves and myocardial cells in the human atrial appendage. *Cell Tiss. Res.* **254**, 155–166

290 Whittaker, V.P. (1982) Biophysical and biochemical studies of isolated cholinergic vesicles from *Torpedo marmorata*. *Fed. Proc.* **41**, 2759–2764

291 Whittaker, V.P. (1984) The structure and function of cholinergic synaptic vesicles. *Biochem. Soc. Transact.* **12,** 561–576

292 Whittaker, V.P. (1988) The synaptic vesicle. In: *Strutcural elements of the nervous system* (Laitha, A., ed.) Plenum, New York, pp.41–96

293 Whittaker, V.P., Michaelson, I.A., Kirkland, R.J.A. (1963) The separation of synaptic vesicles from nerve ending particles. *Biochem. Pharmacol.* **12,** 300–302

294 Whittaker, V.P., Michaelson, I.A., Kirkland, R.J.A. (1964) The separation of synaptic vesicles from nerve-ending particles ('synaptosomes'). *Biochem. J.* **90,** 293–303

295 Wiedenmann, B., Franke, W.W. (1985) Identification and localization of synapto-physin, an integral membrane glycoprotein of Mr 38,000 characteristic of presynaptic vesicles. *Cell* **41,** 1017–1028

296 Wiedenmann, B., Franke, W.W., Kuhn, C., Moll, R., Gould, V.E. (1986) Synaptophysin: A marker protein for neuroendocrine cells and neoplasms. *Proc. Natl. Acad. Sci. USA* **83,** 3500–3504

297 Wiedenmann, B., Rehm, H., Knierim, M., Becker, C.M. (1988) Fractionation of synaptophysin-containing vesicles from rat brain and cultured PC12 pheochromo-cytoma cells. *FEBS Lett.* **240,** 71–77

298 Winkler, H. (1988) Occurrence and mechanism of exocytosis in adrenal medulla and sympathetic nerve. *Handbook of Experimental Pharmacology* **90,** 43–118

299 Winkler, H., Apps, D.K., Fischer-Colbrie, R. (1986) The molecular function of adrenal chromaffin granules: established facts and unresolved topics. *Neuroscience* **18,** 261–290

300 Yamagata, S.K., Parsons, S.M. (1989) Cholinergic synaptic vesicles contain a V-type and a P-type ATPase. *J. Neurochem.* **53,** 1354–1362

301 Yamagata, S.K.; Noremberg, K.; Parsons, S.M. (1989) Purification and subunit composition of a cholinergic synaptic vesicle glycoprotein, phosphointermediate-forming ATPase. *J. Neurochem.* **53,** 1345–1353

302 Yang-Feng, T.L., DeGennaro, L. J., Francke, U. (1986) Genes for synapsin I, a neuronal phosphoprotein, map to conserved regions of human and murine x chromosomes. *Proc. Natl. Acad. Sci. USA* **83,** 8679–8683

303 Zanetta, J.P., Reeber, A., Vincendon, G. (1981) Glycoproteins from adult rat brain synaptic vesicles. Fractionation on four immobilized lectins. *Biochim. Biophys. Acta* **670,** 393–400

304 Zhu, P.C., Thureson-Klein, Å., Klein, R.L. (1986) Exocytosis from large dense cored vesicles outside the active synaptic zones of terminals within the trigeminal subnucleus caudalis: A possible mechanism for neuropeptide release. *Neuroscience* **19,** 43–54

305 Zimmermann, H. (1979) Vesicle recycling and transmitter release. *Neuroscience* **4,** 1773–1804

306 Zimmermann, H. (1982) Isolation of cholinergic nerve vesicles. In: Klein, R., Lagercrantz, H., Zimmermann, H. (eds.) *Neurotransmitter Vesicles.* Academic Press, New York, pp. 241–269.

307 Zimmermann, H. (1988) Cholinergic synaptic vesicles. *Handbook of Experimental Pharmacology* **86,** 349–382

308 Zimniak, L., Dittrich, P., Gogarten, J.P., Kibak, H., Taiz, L. (1988) The cDNA sequence of the 69-kDa subunit of the carrot vacuolar H^+-ATPase. Homology to the β-chain of F_0F_1-ATPases. *J. Biol. Chem.* **263,** 9102

309 Zisapel, N., Zurgil, N. (1979) Studies on synaptic vesicles in mammalian brain. Characterization of highly purified synaptic vesicles from bovine cerebral cortex. *Brain Res.* **178,** 297–310

310 Zisapel, N., Levi, M., Gozes, I. (1980) Tubulin: An integral protein of mammalian synaptic vesicle membranes. *J. Neurochem.* **34,** 26–32

3 Chromogranins/Secretogranins – Widespread Constituents of the Secretory Granule Matrix in Endocrine Cells and Neurons

Wieland B. Huttner, Hans-Hermann Gerdes and *Patrizia Rosa*

3.1 Introduction

In certain specialized eukaryotic cells such as neuroendocrine (NE) cells and neurons, the final step in the secretion of signalling molecules such as peptide hormones, neuropeptides and "classical" neurotransmitters (e.g., acetylcholine) is a highly regulated event. The cellular basis for the regulated release of these molecules is the existence of specific organelles in which these molecules are stored at high concentration and from which they are released when the cell receives an appropriate stimulus. NE cells contain two types of such secretory organelles. These are (a) the small "transparent looking" synaptic vesicles of neurons and their counterparts in endocrine cells, and (b) the larger "dense-cored" secretory granules of endocrine cells and their counterparts in neurons (the so-called large dense core vesicles), which we shall collectively refer to as "NE granules".

NE granules are the organelles used for the storage and release of peptide hormones and neuropeptides. NE granules form in the trans Golgi network (Orci et al. 1987; Tooze et al. 1987; Tooze and Huttner, 1990). Most NE granules contain, in addition to their specific peptide hormones or neuropeptides, one or more of the chromogranin/secretogranin (Cg/Sg) proteins (Rosa et al. 1985a,

---

Abbreviations:

Cg, chromogranin; NE, neuroendocrine; Sg, secretogranin; SDS-PAGE, sodium dodecyl sulfate polyacrylamide gel electrophoresis.

1985b; Winkler et al. 1986; Eiden et al. 1987; Fischer-Colbrie et al. 1987; Huttner et al. 1988). The Cgs/Sgs have a much wider tissue distribution than any individual peptide hormone or neuropeptide and are the most widespread markers for the matrix of NE granules known.

It is the purpose of this article 1. to review the biochemical, cell biological and molecular biological data available for the Cgs/Sgs, and 2. to summarize and compare the immunohistochemical data on these proteins in normal and neoplastic human NE cells and tissues.

3.2 The Chromogranin/Secretogranin Protein Family – Molecular and Cell Biology

The Cgs/Sgs are a family of secretory proteins which share many properties, including a widespread occurrence in NE granules (Rosa et al. 1985b; Fischer-Colbrie et al. 1987). Three proteins have so far been established to belong to the Cg/Sg family: chromogranin A (CgA), chromogranin B (CgB) (also called secretogranin I, SgI) and secretogranin II (SgII), for nomenclature see Eiden et al. 1987). Other acidic proteins such as the 32–35 kd "HISL-19" antigen (Krisch et al. 1986; 1988) and the 21–24 kd protein "7B2" (see Marcinkiewicz et al. 1986, and refs. therein; Martens 1988), which also have a widespread distribution in NE granules and which might turn out to be members of the Cg/Sg protein family in a wider sense, will not be discussed.

The first member of this protein family to be identified was CgA, which was originally described and characterized as the major secretory protein of the bovine adrenal medulla (Helle 1966; Blaschko et al. 1967; Smith and Winkler 1967). It was independently identified as a secretory protein of the parathyroid gland, referred to as secretory protein I (Kemper et al. 1974). Subsequently, it was shown that CgA of the adrenal medulla and secretory protein I of the parathyroid gland were very similar, if not identical, proteins (Cohn et al. 1982).

The second member of the Cg/Sg family to be identified was SgII, which was independently described as an M_r 70,000 sulfated secretory protein (p70) of the bovine anterior pituitary (Rosa and Zanini 1981; Zanini and Rosa 1981) and an M_r 86,000–84,000 tyrosine-sulfated secretory protein (p86/84) of the adrenal medulla-derived cell line PC12 (Huttner and Lee 1982; Lee and Huttner 1983). Subsequently, it was shown that a protein immunologically related to p70 was present in the adrenal medulla and that this protein as well as p86/84 of PC12 cells were the same as p70 of the anterior pituitary (Rosa and Zanini 1983; Rosa et al. 1985b). SgII has previously been called TSP 86/84 (Rosa et al. 1985a), GP-87 (Chanat et al. 1986, 1988) or chromogranin C (Fischer-Colbrie et al. 1987).

The third member of the Cg/Sg family to be described was CgB (SgI), which was first characterized as an M_r 113,000–105,000 tyrosine-sulfated secretory protein (p113/p105) of PC12 cells (Huttner and Lee 1982, Lee and Huttner 1983). This protein was subsequently detected in the bovine adrenal medulla (Falkensammer et al. 1985).

Following their initial characterization, the widespread distribution of CgA (Lloyd and Wilson 1983; O'Connor et al. 1983a, 1983b; Cohn et al. 1984; Somogyi et al. 1984), SgII (Rosa et al. 1985a, 1985b) and CgB (SgI) (Fischer-Colbrie et al. 1985; Rosa et al. 1985b) in peptide-secreting NE cells was established, as is described in detail below (see also refs. in Tabs. 3-2 and 3-3). CgA- and CgB-like proteins have been detected in various vertebrate and invertebrate species (Rieker et al. 1988).

3.2.1 Structure and Biochemical Properties

Over the past four years, the primary structure of bovine, human, porcine and rat CgA (Benedum et al. 1986; Iacangelo et al. 1986; Ahn et al. 1987; Konecki et al. 1987; Helman et al. 1988a; Iacangelo et al. 1988a, 1988b; Hutton et al. 1988a; Parmer et al. 1989), of human and rat CgB (SgI), (Benedum et al. 1987; Forss-Petter et al. 1989) and of rat and human SgII (Gerdes et al. 1988, 1989) has been deduced from the corresponding cDNA sequences (Fig. 3-1). Much of the information about the three proteins revealed by these and other biochemical studies is listed in Tab. 3-1. Except for the cleaved signal peptides, the Cgs/Sgs are very hydrophilic proteins with a high proportion of charged, mostly acidic, amino acid residues. These structural features contribute to the remarkable property of all three proteins to remain soluble after boiling (Huttner and Lee 1982; Rosa et al. 1985b). All three proteins are characterized by very low isoelectric points, with CgA (pI $\approx$ 4.8) being the most acidic and CgB (pI $\approx$ 5.2) being the least acidic polypeptide. These low isoelectric points are not only due to the high proportions of acidic amino acid residues but also result from post-translational modifications such as phosphorylation, sialylation and sulfation (Fischer-Colbrie et al. 1982; Kiang et al. 1982; Lee and Huttner 1983; Rosa et al. 1985b, and refs. therein).

An unexpected finding of the cDNA cloning work was that the actual polypeptide length of these proteins, particularly in the case of CgA (Benedum et al. 1986; Iacangelo et al. 1986) and CgB (Benedum et al. 1987), is considerably less than their apparent M_r on SDS-PAGE. The anomalous

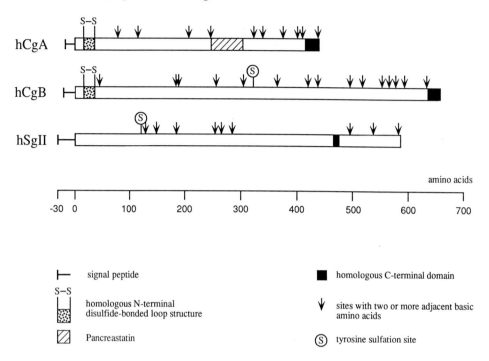

Fig. 3–1 Schematic diagram of the primary structure of human chromogranin A (hCgA), chromogranin B (hCgB) and secretogranin II (hSgII).

electrophoretic mobility of the Cgs/Sgs is, for the most part, not due to post-translational modifications but reflects an inherent property of the unmodified polypeptides, presumably the abundance of acidic amino acid residues, as observed for certain other acidic proteins, e.g., the neurofilament polypeptides (Kaufmann et al. 1984).

The sequences of the Cgs/Sgs further revealed that, although all three proteins share several common properties, CgA and CgB are structurally more closely related to each other than either protein is to SgII. CgA and CgB contain a highly homologous, disulfide-bonded loop structure near their N-termini and another homologous sequence at their C-termini (Benedum et al. 1987). The disulfide-bonded loop structure is not found in SgII, which lacks cysteine residues, whereas a significant homology to the C-terminal sequence of CgA and CgB is present in SgII (Gerdes et al. 1988, 1989).

Finally, we know about two other structural and biochemical properties of the Cgs/Sgs which will be discussed in detail below as they are probably related to their function(s). First, all three proteins contain multiple dibasic residues which are potential sites for the proteolytic processing to smaller, perhaps biologically active, peptides. Secondly, these proteins bind calcium with

Tab. 3-1 Properties of the chromogranins/secretogranins.

Properties	Chromogranin A	Chromogranin B	Secretogranin II
M_r in SDS-PAGE	75,000–85,000	100,000–120,000	84,000–87,000
Mol. wt. (cDNA-deduced)	48,000	76,000	67,000
Isoelectric point	5.0–4.8	5.2–5.0	5.1–4.9
Heat stability	yes	yes	yes
Signal peptide	yes	yes	yes
Subcellular localization	NE granules	NE granules	NE granules
Secreted	yes	yes	yes
Calcium binding	yes	yes	yes
affinity	high μM	ND	ND
number of sites	many	ND	ND
aggregation	yes	yes	yes
Post-translational modifications			
N-glycosylation	no	no	no
O-glycosylation	yes	yes	yes
GAG chains	yes	ND	ND
Phosphorylation	Ser, Thr	Ser, Thr	Ser, Thr
Sulfation	Tyr*, carbohydrate	Tyr, carbohydrate	Tyr, carbohydrate
Proteolytic processing	yes	yes	yes
α-amidation	yes (pancreastatin)	ND	ND
Cellular distribution			
endocrine cells	yes	yes	yes
neurons	yes	yes	yes
exocrine cells	no	no	no
constitutive cells	no	no	no

For refs., see text. ND, not determined. *Only in some species.

moderate affinity at multiple sites (Reiffen and Gratzl 1986; Gorr et al. 1988; Cozzi and Zanini 1988) and aggregate in the presence of this divalent cation (Gorr et al. 1988; Gerdes et al. 1989; Gorr et al. 1989). These properties are probably a reflection of both the excess negative charge and the presumptive secondary structure of these proteins, which largely consist of helices and turns, allowing intra- as well as intermolecular calcium binding.

3.2.2 Subcellular Localization

3.2.2.1 Immunocytochemistry

Immunoelectron microscopic data show that the Cgs/Sgs are present in the matrix of NE granules (Rosa et al. 1985a; Varndell et al. 1985; Ehrhart et al. 1986; Hashimoto et al. 1987; Hearn 1987; Silver et al. 1988; Tougard et al. 1989; Steiner et al. 1989; Bassetti et al. 1990) which are also known to contain peptide hormones and neuropeptides. In some cases, co-localization of the Cgs/Sgs with specific hormones in NE granules has been demonstrated directly by double immunoelectron microscopy (Ehrhart et al. 1986; Hashimoto et al. 1987; Silver et al. 1988; Steiner et al. 1989; Bassetti et al. 1990). Immuno-reactivity for the Cgs/Sgs has also been detected in the Golgi complex (Hashimoto et al. 1987; Tougard et al. 1989).

3.2.2.2 Subcellular Fractionation

Consistent with the results of immunocytochemistry, all three Cgs/Sgs are largely recovered in subcellular fractions enriched in NE granules, as exemplified for chromaffin granules (Rosa et al. 1985b; Winkler et al. 1986).

3.2.3 Biosynthesis and Proteolytic Processing

The relative proportion of the individual members of the Cg/Sg family in a given tissue (e.g., the adrenal medulla) differs across species, and in a given species (e.g., cow) differs across tissues (Rosa et al. 1985b; for review see Fischer-Colbrie et al. 1987). This suggests that the synthesis of the individual Cgs/Sgs is differentially regulated. Indeed, different protein and mRNA levels for CgA and B after neurogenic and humoral stimulation have been reported (Sietzen et al. 1987; Fischer-Colbrie et al. 1988). Glucocorticoids increase the amount of CgA mRNA and protein, but do not appear to affect the levels of CgB and SgII proteins (Rausch et al. 1988; Grino et al. 1989; Fischer-Colbrie et al. 1989). Phorbol esters also increase the CgA mRNA level and the synthesis of the protein (Murray et al. 1988; Simon et al. 1989); the detailed mechanisms underlying these events remain to be established.

After translation, translocation into the lumen of the rough endoplasmic reticulum and signal peptide cleavage, the Cgs/Sgs undergo a variety of covalent modifications during their passage from the rough endoplasmic reticulum to the trans Golgi network, including formation of a single disulfide bond (only CgA and CgB), O-glycosylation, phosphorylation (largely on serine) and sulfation of carbohydrate and tyrosine residues (Kiang et al. 1982; Rosa et al. 1985b; Benedum et al. 1987; Fischer-Colbrie et al. 1987, and refs. therein; see Tab. 3-1). In the trans Golgi network, the Cgs/Sgs are packaged into NE granules (Tooze and Huttner, 1990). Secretory proteins that normally are not packaged into NE granules are re-routed to these organelles if complexed to the Cgs/Sgs (Rosa et al. 1989). In NE granules, the Cgs/Sgs undergo partial proteolytic processing during their storage (Rosa et al. 1985b; Fischer-Colbrie et al. 1987; Hutton et al. 1987a, 1987b; Wohlfarter et al. 1988). Defined proteolytic products of the Cgs/Sgs include betagranin (Hutton et al. 1987a; 1987b; 1988a) and pancreastatin (Tatemoto et al. 1986; Eiden 1987; Huttner and Benedum 1987; Konecki et al. 1987; Iacangelo et al. 1988b; Sekiya et al. 1988; Schmidt et al. 1988b; Funakoshi et al. 1989) in the case of CgA, and GAWK in the case of CgB (Benjannet et al. 1987). Upon stimulation of exocytosis, these proteins and their proteolytic fragments are released into the extracellular space (Blaschko et al. 1967; Rosa et al. 1985b). CgA and CgB have been detected in the blood plasma (O'Connor and Bernstein 1984; O'Connor and Deftos 1986; Wiedenmann et al. 1987).

3.2.4 Functions

The physiological function of the Cgs/Sgs is still an open question. The Cgs/Sgs may have extracellular roles in that the intact proteins or proteolytic fragments derived from them exert biological activities on target cells (Eiden 1987; Huttner and Benedum 1987; Simon et al. 1988). The determination of the primary structure of CgA, CgB and SgII has shown that all three proteins contain multiple dibasic sites for potential proteolytic processing (Benedum et al. 1986, 1987; Iacangelo et al. 1986; Gerdes et al. 1988, 1989). CgA is the precursor of pancreastatin, a peptide which partially inhibits glucose-induced insulin release from the isolated pancreas and regulated protein secretion from certain other cells (Tatemoto et al. 1986; Eiden 1987; Huttner and Benedum 1987; Konecki et al. 1987; Iacangelo et al. 1988b; Sekiya et al. 1988; Schmidt et al. 1988b; Fasciotto et al. 1989). In addition, tryptic fragments of CgA partially inhibit regulated exocytosis from isolated chromaffin cells (Simon et al. 1988). On the other hand, full-length CgA appears to exert similar inhibitory actions as pancreastatin (Greeley et al. 1989). Thus, it is possible that CgA and its proteolytic products function as autocrine, paracrine or even endocrine regulatory agents on secretory processes, although it remains to be established whether the above *in vitro* observations reflect the true functions of the Cgs/Sgs *in vivo*.

It has also been suggested that the Cgs/Sgs have intracellular roles in the packaging and/or processing of certain peptide hormones and neuropeptides (Rosa et al. 1985a, 1985b; Seidah et al. 1987). It is conceivable that the low pH- and Ca^{2+}-promoted aggregation of the Cgs/Sgs may be relevant in this regard (Gerdes et al. 1989; Gorr et al. 1989). In this context, it is interesting that in the bovine anterior pituitary, thyroid-stimulating hormone and luteinizing hormone, which are co-packaged together with the Cgs/Sgs into NE granules of thyrotrophs and gonadotrophs, are also co-packaged with these proteins into a subpopulation of NE granules of somatomammotrophs that are distinct from another NE granule subpopulation containing growth hormone and prolactin (Bassetti et al. 1990). These observations are consistent with the possibility that the Cgs/Sgs and certain peptide hormones co-aggregate in the packaging process.

3.3 The Chromogranin/Secretogranin Protein Family – Markers for Neuroendocrine Cells

3.3.1 Cellular Distribution in Normal Cells and Tissues

Prior to describing the cellular distribution of the Cgs/Sgs in normal and neoplastic cells and tissues, we would like to discuss possible causes for an apparent lack or variability of immunoreactivity in certain immunohistochemical preparations. The Cgs/Sgs (a) vary greatly in abundance between different cell types and different NE granules, (b) may be complexed to other molecules *in situ* (c) exhibit marked sequence variations across species, and (d) undergo extensive post-translational modifications (see Tab. 3-1). These features may contribute to a reduced immunoreactivity of the Cgs/Sgs, which should be borne in mind in the interpretation of negative or variable immunohistochemical results.

3.3.1.1 Animal Cells and Tissues

The cellular distribution of CgA, CgB and SgII in normal animal NE cells, as revealed by immunohistochemical studies from many laboratories, is presented in Tab. 3-2. For a given tissue, the cellular distribution of these three proteins is sometimes overlapping and sometimes complementary (Rosa et al. 1985b; Rindi et al. 1986; Fischer-Colbrie et al. 1987). The latter finding is particularly important since it implies that lack of immunoreactivity for a specific protein may not necessarily mean the absence of all members of this protein family in general. Thus, the use of a cocktail of antibodies against all three Cgs/Sgs appears to be the method of choice to identify NE granule-containing cells in the future. However, even then it cannot be excluded at present that certain NE granule-containing cells may lack immunoreactivity.

3.3.1.1.1 Nervous System

All three Cgs/Sgs have been detected in neurons (see Tab. 3-2). In fact, these proteins are more widespread markers of NE granules of neurons than any known neuropeptide. However, as exemplified for the cerebellar cortex, the immunoreactivity for SgII (Cozzi et al. 1989) is more restricted than that for markers of small NE vesicles, e.g. synaptophysin (Navone et al. 1986). This most probably reflects the more widespread occurrence of small NE vesicles than NE granules in neuronal tissue.

3.3.1.1.2 Endocrine System

Many endocrine cells have been found to be strongly immunoreactive for Cgs/Sgs. These include endocrine cells of the anterior pituitary, C-cells of the thyroid, chief-cells of the parathyroid, chromaffin cells of the adrenal medulla, various islet cells of the pancreas, NE cells of the bronchial and gastrointestinal tract, of lymph nodes and of the thymus, as well as Merkel cells of the skin (see Tab. 3-2).

Many Cg/Sg-positive NE cells contain all three of these proteins; however, as originally noted (Rosa et al. 1985b), several NE cell types have been found that appear to express only one or two of them. For example, C-cells of the thyroid contain CgA and SgII but appear to lack CgB immunoreactivity (Rindi et al. 1986); chief-cells of the parathyroid contain CgA but appear to lack CgB and SgII immunoreactivity (Lassmann et al. 1986).

Recently, antibodies directed against defined CgA fragments such as betagranin (Hutton et al. 1988b) and pancreastatin (Schmidt et al. 1988a) or against defined CgB fragments such as GAWK (Iguchi et al. 1988; Salahuddin et al. 1989) have also been used to investigate the distribution of these Cg/Sg products in NE cells.

Tab. 3-2 Immunohistochemical identification of chromogranins/secretogranins in normal animal NE cells.

Cells	Chromogranin A	Chromogranin B	Secretogranin II
Neurons	++ Cohn et al. 1984 ++ Somogyi et al. 1984 ++ Nolan et al. 1985 ++ Lassmann et al. 1986 ++ Volknandt et al. 1987	++ Rosa et al. 1985b ++ Fischer-Colbrie et al. 1985 ++ Lassmann et al. 1986	++ Rosa et al. 1985b ++ Cozzi et al. 1989 ++ Lassmann et al. 1986
Endocrine cells of the anterior pituitary			
ACTH	− O'Connor et al. 1983a − Cohn et al. 1984 + Rundle et al. 1986	+ Rundle et al. 1986	++ Rosa et al. 1985a − Rundle et al. 1986
FSH/LH	+ O'Connor et al. 1983a + Rundle et al. 1986 ++ Rindi et al. 1986	++ Rindi et al. 1986 ++ Rundle et al. 1986	++ Rosa et al. 1985a ++ Rindi et al. 1986 ++ Rundle et al. 1986
GH	+ O'Connor et al. 1983a − Cohn et al. 1984 − Rundle et al. 1986	− Rundle et al. 1986	− Rosa et al. 1985a − Rundle et al. 1986 ++ Hashimoto et al. 1987
Prolactin	− O'Connor et al. 1983a − Cohn et al. 1984 − Rundle et al. 1986	− Rundle et al. 1986	++ Rosa et al. 1985a − Rundle et al. 1986 ++ Hashimoto et al. 1987
TSH	++ Cohn et al. 1984 ++ Rindi et al. 1986 ++ Rundle et al. 1986	++ Rindi et al. 1986 − Rundle et al. 1986	++ Rosa et al. 1985a ++ Rindi et al. 1986 + Rundle et al. 1986

Cell type	Column 1	Column 2	Column 3
Unspecified	++ O'Connor et al. 1983a ++ Cohn et al. 1984 ++ Somogyi et al. 1984 + Nolan et al. 1985 ++ Lassmann et al. 1986	++ Rosa et al. 1985b + Fischer-Colbrie et al. 1985 ++ Lassmann et al. 1986	++ Rosa et al. 1985a ++ Rosa et al. 1985b ++ Lassmann et al. 1986
C-cells of the thyroid	++ O'Connor et al. 1983a ++ Cohn et al. 1984 ++ Nolan et al. 1985 – Lassmann et al. 1986 ++ Rindi et al. 1986	– Lassmann et al. 1986 – Rindi et al. 1986	++ Rosa et al. 1985b – Lassmann et al. 1986 ++ Rindi et al. 1986
Chief-cells of the parathyroid	+ O'Connor et al. 1983a – Cohn et al. 1984 ++ Nolan et al. 1985 ++ Lassmann et al. 1986 ++ Rindi et al. 1986	+ Fischer-Colbrie et al. 1985 – Lassmann et al. 1986 – Rindi et al. 1986	– Lassmann et al. 1986 + Rindi et al. 1986
Chromaffin cells of the adrenal medulla	++ O'Connor et al. 1983a ++ Cohn et al. 1984 ++ Somogyi et al. 1984 ++ Nolan et al. 1985 ++ Lassmann et al. 1986 ++ Yoshie et al. 1987	++ Rosa et al. 1985b ++ Fischer-Colbrie et al. 1985 ++ Lassmann et al. 1986 ++ Rindi et al. 1986 ++ Yoshie et al. 1987	++ Rosa et al. 1985b ++ Lassmann et al. 1986 ++ Rindi et al. 1986 + Yoshie et al. 1987
Islet cells of the pancreas **A cells**	++ Ehrhart et al. 1986 ++ Grube et al. 1986 ++ Rindi et al. 1986 ++ Yoshie et al. 1987 ++ Ehrhart et al. 1988 + Buffa et al. 1988a	+ Rindi et al. 1986 – Yoshie et al. 1987 + Buffa et al. 1988a	++ Rindi et al. 1986 + Yoshie et al. 1987 + Buffa et al. 1988a

Tab. 3-2 (continued)

Cells	Chromogranin A	Chromogranin B	Secretogranin II
B cells	++ O'Connor et al. 1983a ++ Ehrhart et al. 1986 ++ Grube et al. 1986 + Rindi et al. 1986 ++ Yoshie et al. 1987 ++ Ehrhart et al. 1988	− Rindi et al. 1986 − Yoshie et al. 1987	− Rindi et al. 1986 + Yoshie et al. 1987
D cells	+ Ehrhart et al. 1986 + Grube et al. 1986 + Yoshie et al. 1987 + Ehrhart et al. 1988	− Yoshie et al. 1987	− Yoshie et al. 1987
PP cells	+ Ehrhart et al. 1986 + Grube et al. 1986 ++ Rindi et al. 1986 + Yoshie et al. 1987 + Ehrhart et al. 1988 + Buffa et al. 1988a	− Rindi et al. 1986 − Yoshie et al. 1987	+ Rindi et al. 1986 ++ Yoshie et al. 1987
Unspecified	++ O'Connor et al. 1983a − Cohn et al. 1984 ++ Nolan et al. 1985 ++ Lassmann et al. 1986	− Rosa et al. 1985b ++ Fischer-Colbrie et al. 1985 + Lassmann et al. 1986	+ Rosa et al. 1985b + Lassmann et al. 1986

NE-cells of the

Bronchial tract	− ++ *	Nolan et al. 1985 Lauweryns et al. 1987	ND		ND	
Gastrointestinal tract	++ ++ ++ ++ ++	Cohn et al. 1984 Nolan et al. 1985 Lassmann et al. 1986 Rindi et al. 1986 Buffa et al. 1988a	++ ++ ++ ++ +	Rosa et al. 1985b Fischer-Colbrie et al. 1985 Lassmann et al. 1986 Rindi et al. 1986 Buffa et al. 1988a	+ − − +	Rosa et al. 1985b Lassmann et al. 1986 Rindi et al. 1986 Buffa et al. 1988a
Lymph nodes and thymus	++ +/−	Hogue-Angeletti & Hickey, 1985 Nolan et al. 1985	ND		ND	
Merkel cells of the skin	− ++ ++ ++	Nolan et al. 1985 Gauweiler et al. 1988 Hartschuh & Weihe 1988 Hartschuh et al. 1989	ND		ND	

++   all or the majority of NE cells reported to be immunoreactive
+   the minority of NE cells reported to be immunoreactive
−   no detectable immunoreactivity
ND   Not determined
*   Commercially available monoclonal antibody LK2H10 used to detect immunoreactive chromogranin A
Note that the results obtained for several animal species are compiled in this Table, which explains some of the apparent discrepancies.

3.3.1.2 Human NE Cells and Tissues

The cellular distribution of CgA, CgB and SgII in normal human NE cells, as revealed by immunohistochemical studies from many laboratories, is presented in Tab. 3-3. For CgA, many of these studies have been performed using the monoclonal antibody LK2H10 described by Lloyd and Wilson (1983) which is commercially available (e.g., from Camon, Wiesbaden, F.R.G). For CgB, so far only one study using monoclonal antibodies has been reported (Pelagi et al. 1989).

3.3.1.2.1 Nervous System

So far, only a few studies have been published on the Cgs/Sgs in normal human nervous system (see Tab. 3-3). These show that CgA and CgB are present in human neurons.

3.3.1.2.2 Endocrine System

A variety of normal human endocrine cells have been shown to contain members of the Cg/Sg protein family. These include endocrine cells of the anterior pituitary, C-cells of the thyroid, chief-cells of the parathyroid, chromaffin cells of the adrenal medulla, islet cells of the pancreas, NE cells of the bronchial and gastrointestinal tract and Merkel cells of the skin (Tab. 3-3). Most of these studies have been concerned with CgA, using the monoclonal antibody LK2H10 (see asterisks in Tab. 3-3). Data on the cellular distribution of CgB and SgII are, so far, scarcer. Recently, an antibody against a defined CgB fragment, GAWK, has been used to study the distribution of this protein in human NE tissues (Bishop et al. 1989).

3.3.2 Occurrence in Neoplastic Human Cells and Tissues

The occurrence of CgA, CgB and SgII in human NE tumors, as revealed by immunohistochemical studies from many laboratories, is presented in Tab. 3-4.

3.3.2.1 Nervous System-Derived Tumors

Only neuroblastomas and paragangliomas have so far been studied for the occurrence of Cgs/Sgs. Whereas most paragangliomas were found to be immunoreactive for CgA and CgB by most investigators (see Tab. 3-4; see also, however, Scheithauer et al. 1986), only a few neuroblastomas showed a – mostly weak – immunoreactivity with antibody LK2H10 for CgA. No immunoreactivity in neuroblastomas was observed using antibodies against rat CgB (Lloyd et al. 1988). Studies on the occurrence of SgII in human nervous-system derived NE tumors have not yet been reported.

3.3.2.2 Endocrine System-Derived Tumors

Many NE tumors derived from the endocrine system have been shown to contain Cgs/Sgs (see Table 3-4). Most studies have focussed on CgA, using the monoclonal antibody LK2H10.

Pituitary adenomas have only been studied with antibodies against CgA and CgB (see Tab. 3.4). Whereas prolactinomas showed no immunostaining with antibody LK2H10, these adenomas could be immunostained with polyclonal antibodies against CgB (Lloyd et al. 1988). So far, no data exist on the expression of SgII in pituitary adenomas.

Pheochromocytomas and medullary thyroid carcinomas have been shown to express the Cgs/Sgs uniformly (see Tab. 3-4).

For gastrointestinal carcinoids and NE carcinomas, CgA has been shown (with the possible exceptions of some hind-gut carcinoids) to be a very reliable marker (see Tab. 3-4). While these tumors appear to express large amounts of CgA, they seem to contain lesser amounts of CgB and SgII, showing weaker immunohistochemical reactions for the latter two proteins (Weiler et al. 1987; Lloyd et al. 1988; Wiedenmann et al. 1988).

Pancreatic islet cell tumors have been studied for the expression of the Cgs/Sgs in detail. In contrast to earlier findings (Lloyd et al. 1984), it is now well accepted that CgA is well detectable in pancreatic insulinomas as well as in pancreatic glucagonomas, gastrinomas and hormone-inactive tumors (see Tab. 3-4). Concerning the expression of CgB and SgII in islet cell tumors, all neoplasias studied, with the exception of gastrinomas (which were found to lack CgB immnuoreactivity), showed immunoreactivity with antibodies against these two proteins (Wiedenmann et al. 1988).

In contrast to the above mentioned tumors, small cell carcinomas of the lung have been shown to express only CgA, and only in a low percentage of cases (see Tab. 3-4). While two studies (Lloyd et al. 1988; Weiler et al. 1988) reported

Tab. 3-3 Immunohistochemical identification of chromogranins/secretogranins in normal human NE cells.

Cells	Chromogranin A	Chromogranin B	Secretogranin II
Neurons	− * Lloyd & Wilson 1983 − * Wilson & Lloyd 1984 ++* Lloyd et al. 1986	++ Lloyd et al. 1988 ++ Schmid et al. 1989	ND
Endocrine cells of the anterior pituitary			
ACTH	−* Lloyd et al. 1985a	ND	ND.
FSH/LH	++* Lloyd et al. 1985a	ND	ND
GH	−* Lloyd et al. 1985a	ND	ND
Prolactin	−* Lloyd et al. 1985a	ND	ND
TSH	++* Lloyd et al. 1985a	ND	ND
Unspecified	++* Lloyd & Wilson 1983 ++ O'Connor et al. 1983b ++* Wilson & Lloyd 1984	++ Hagn et al. 1986 ++ Lloyd et al. 1988 + Schmid et al. 1989	++ Hagn et al. 1986
C-cells of the thyroid	++ Lloyd & Wilson 1983 ++ O'Connor et al. 1983b ++* Wilson & Lloyd 1984 ++ Rindi et al. 1986 ++* Lauweryns et al. 1987	++ Schmid et al 1989	ND

Chief-cells of the parathyroid	++* Lloyd & Wilson 1983 ++ O'Connor et al. 1983b ++ Wilson & Lloyd 1984 ++ Hagn et al. 1986 ++ Rindi et al. 1986 ++ Hearn, 1987	++ Lloyd et al. 1988 - Schmid et al. 1989	ND
Chromaffin cells of the adrenal medulla	++* Lloyd & Wilson 1983 ++ O'Connor et al. 1983b ++* Wilson & Lloyd 1984 ++* Lloyd et al. 1985b ++ Hagn et al. 1986 ++* Lloyd et al. 1986 ++ Schober et al. 1987 ++* Wiedenmann et al. 1988 ++ Lloyd et al. 1988	++ Hagn et al. 1986 ++ Schober et al. 1987 ++ Lloyd et al. 1988 ++ Schmid et al. 1989	++ Hagn et al. 1986 ++ Schober et al. 1987
Islet cells of the pancreas			
A cells	++ O'Connor et al. 1983b ++* Lloyd et al. 1984 -* Varndell et al. 1985 + Grube et al. 1986 ++ Rindi et al. 1986 +* Wiedenmann et al. 1988	++ Rindi et al. 1986 + Lloyd et al. 1988 + Wiedenmann et al. 1988 ++ Schmid et al. 1989	+ Rindi et al. 1986 + Wiedenmann et al. 1988
B cells	++ O'Connor et al. 1983b +* Lloyd et al. 1984 -* Varndell et al. 1985 + Grube et al. 1986 +* Wiedenmann et al. 1988	ND	ND
D cells	++ O'Connor et al. 1983b ++* Varndell et al. 1985 - Grube et al. 1986 +* Wiedenmann et al. 1988	ND	ND

Tab. 3-3 (continued)

Cells	Chromogranin A	Chromogranin B	Secretogranin II
PP cells	−* Varndell et al. 1985 − Grube et al. 1986 ++* Rindi et al. 1986	+ Schmid et al. 1989	ND
Unspecified	++* Lloyd & Wilson 1983 ++ O'Connor et al. 1983b ++* Wilson & Lloyd 1984 ++ Hagn et al. 1986 ++* Lloyd et al. 1988 +* Wiedenmann et al. 1988	+ Hagn et al. 1986 + Lloyd et al. 1988 + Wiedenmann et al. 1988	− Hagn et al. 1986 + Wiedenmann et al. 1988
NE-cells of the			
Bronchial tract	++ Wilson & Lloyd 1984 ++ * Lauweryns et al. 1987	ND	ND
Gastrointestinal tract	++* Lloyd & Wilson 1983 ++ O'Connor et al. 1983b ++* Wilson & Lloyd 1984 ++* Facer et al. 1985 ++* Varndell et al. 1985 ++ Rindi et al. 1986 ++ Pelagi et al. 1989	+ Lloyd et al. 1988 + Wiedenmann et al. 1988 ++ Pelagi et al. 1989 ++ Schmid et al. 1989	+ Wiedenmann et al. 1988
Merkel cells of the skin	− * Wilson & Lloyd 1984 ++ * Hartschuh & Weihe 1988 ++ * Hartschuh et al. 1989	ND	ND

++   all or the majority of NE cells reported to be immunoreactive
+   the minority of NE cells reported to be immunoreactive
−   no detectable immunoreactivity
ND   Not determined

Tab. 3-4 Immunohistochemical identification of chromogranins/secretogranins in human NE tumors.

Tumor	Chromogranin A	Chromogranin B	Secretogranin II
Adrenal Medulla			
Pheochromocytoma	++ * 3/3 O'Connor et al. 1983b ++ * 5/5 Lloyd & Wilson 1983 ++ *25/25 Wilson & Lloyd 1984 ++ *28/28 Lloyd et al. 1985b ++* 5/5 Varndell et al. 1985 ++ NS Schober et al. 1987 ++ 36/36 Kimura et al. 1988 ++ NS Pelagi et al. 1989	++ 4/4 Lloyd et al. 1988 ++ 2/2 Schober et al. 1987 ++ 2/2 Wiedenmann et al. 1988 ++ NS Pelagi et al. 1989	++ 2/2 Schober et al. 1987 ++ 2/2 Wiedenmann et al. 1988
Breast			
Agyrophilic carcinoma	+ * 2/9 Bussolati et al. 1985 + * 1/1 Azzopardi et al. 1986 + 3/3 Nesland et al. 1986 ++ *8/20 Bussolati et al. 1987	ND	ND
Gastrointestinal Tract			
Stomach Carcinoid	+ * 3/3 Wilson & Lloyd 1984 ++ * 4/6 Müller et al. 1987 ++ * 3/4 Kimura et al. 1988 + * 1/1 Wiedenmann et al. 1988	+ 1/1 Wiedenmann et al. 1988	+ 1/1 Wiedenmann et al. 1988
Adenocarcinoma	+ *28/150 Ooi et al. 1988	ND	ND
Composite carcinoma	+ * 1/1 Ulich et al. 1988	ND	ND

Tab. 3-4 (continued)

Tumor	Chromogranin A	Chromogranin B	Secretogranin II
Small intestine			
Carcinoid	++ * 2/2 Wilson & Lloyd 1984	+ 12/13 Weiler et al. 1987	+ 15/15 Wiedenmann et al. 1988
	++ * 9/9 Dayal et al. 1986	+ 2/2 Lloyd et al. 1988	+ 8/13 Weiler et al. 1987
	++ * 9/11 Nash & Said 1986	+ 15/15 Wiedenmann et al. 1988	
	++ * 5/5 Hearn 1987		
	++ * 2/2 Lloyd et al. 1988		
	++ *15/15 Wiedenmann et al. 1988		
NE carcinoma	ND	ND	ND
Large intestine			
Carcinoid	++ * 2/5 Nash & Said 1986	+ 1/1 Wiedenmann et al. 1988	+ 1/1 Wiedenmann et al. 1988
	++ 6/6 Bishop et al. 1988		
	++ 24/27 Kimura et al. 1988		
	++ * 1/1 Wiedenmann et al. 1988		
NE carcinoma	++ * 3/10 Wick et al. 1987	ND	ND
Liver			
Carcinoid	+ 1/1 Miura and Shirasawa 1988	ND	ND
Lung			
Carcinoid	++ * 9/9 Said et al. 1985	+ 14/14 Weiler et al. 1987	+ 14/14 Weiler et al. 1987
	++ * 6/6 Walts et al. 1985	+ 2/2 Lloyd et al. 1988	
	++ 14/14 Weiler et al. 1987		
	++ 10/10 Bishop et al. 1988		
	++ * 4/4 Linnoila et al. 1988		
Small cell carcinoma	+ * 4/10 Wilson & Lloyd 1984	− 0/4 Lloyd et al. 1988	− 0/1 Weiler et al. 1988
	− * 0/12 Said et al. 1985	− 0/1 Weiler et al. 1988	
	− * 0/12 Walts et al. 1985		
	− 0/10 Bishop et al. 1988		
	− * 0/4 Lloyd et al. 1988		
	+ * 3/8 Buffa et al. 1988b		
	+ * 1/1 Weiler et al. 1988		
	+ * 18/32 Linnoila et al. 1988		

Tumor						
NE large cell carcinoma	++ * 3/11	Mooi et al. 1988	ND		ND	
	++ 1/52	Piehl et al. 1988				
	++ * 2/15	Linnoila et al. 1988				
Middle ear						
Carcinoid	++ * 1/1	McNutt & Bolen 1985	ND		ND	
	++ 1/1	Kimura et al. 1988				
Nervous system						
Ganglioneuroblastoma	ND		ND		ND	
Ganglioneuroma	ND		ND		ND	
Medulloblastoma	ND		ND		ND	
Neuroblastoma	+ * 4/10	Wilson & Lloyd 1984	– 0/2	Lloyd et al. 1988	ND	
	+ * 2/6	Lloyd et al. 1986				
Paraganglioma	++ * 5/5	Wilson & Lloyd 1984	+ 3/3	Lloyd et al. 1988	ND	
	++ * 4/4	Johnson et al. 1985				
	++ * 9/9	Lloyd et al. 1986				
	– * 0/11	Scheithauer et al. 1986				
	++ * 26/29	Johnson et al. 1988				
Ovary						
Carcinoid	++ * 1/1	Lloyd & Wilson 1983	ND		ND	
	++ * 5/6	Stagno et al. 1987				
	++ 2/2	Bishop et al. 1988				
	++ 2/2	Kimura et al. 1988				
Pancreas						
Gastrinoma	++ * 11/11	Lloyd et al. 1984	– 0/5	Wiedenmann et al. 1988	+ 5/5	Wiedenmann et al. 1988
	++ * 3/3	Varndell et al. 1985				
	++ * 2/3	Chejfec et al. 1987				
	++ * 5/5	Wiedenmann et al. 1988				
	++ * 2/2	Bordi et al. 1988				
Glucagonoma	++ * 3/3	Lloyd et al. 1984	ND		ND	
	++ * 7/10	Hamid et al. 1986				
	++ * 3/4	Chejfec et al. 1987				
	++ * 15/15	Bordi et al. 1988				

Tab. 3-4 (continued)

Tumor	Chromogranin A	Chromogranin B	Secretogranin II
Insulinoma	− * 0/5 Lloyd et al. 1984 + * 7/15 Chejfec et al. 1987 + * 1/3 Buffa et al. 1988b ++ * 5/7 Wiedenmann et al. 1988 + * 15/15 Bordi et al. 1988	− 0/4 Lloyd et al. 1988 + 7/7 Wiedenmann et al. 1988	+ 7/7 Wiedenmann et al. 1988
PP-oma	++ *14/14 Bordi et al. 1988		
Somatostatinoma	++ * 2/2 Bishop et al. 1988 + * 1/1 Buffa et al. 1988b	ND	ND
VIP-oma	− * 0/2 Chejfec et al. 1987 + * 3/6 Buffa et al. 1988b ++ * 1/1 Bordi et al. 1988	ND	ND
Unspecified NE tumors	++ * 1/1 O'Connor et al. 1983b ++* 5/5 Lloyd & Wilson 1983 ++* 15/20 Wilson and Lloyd 1984 ++ * 3/3 Walts et al. 1985	ND	ND
NE tumors without clinical hormonal syndrome	++ * 4/5 Chejfec et al. 1987 + * 6/8 Bordi et al. 1988 ++ 1/1 Sobol et al. 1989	ND	ND
Parathyroid Adenoma	+ 1/1 O'Connor et al. 1983b + * 2/2 Lloyd & Wilson 1983 + * 2/3 Wilson & Lloyd 1984 + * 3/3 Lloyd et al. 1988 + 27/134 Oka et al. 1988	+ 3/3 Lloyd et al. 1988 − 0/1 Weiler et al. 1988	− 0/1 Weiler et al. 1988
Carcinoma	+ * 1/1 Murphy et al. 1986	ND	ND

Pituitary (anterior)
Adenomas

ACTH	++	*	6/8	DeStephano et al. 1984	ND		ND
	–	*	0/14	Lloyd et al. 1985a			
	++	*	1/2	Buffa et al. 1988b			
	+		1/6	Stefaneanu et al. 1988			
FSH/LH	++	*	1/1	DeStephano et al. 1984	+	1/1	ND
	+	*	6/9	Lloyd et al. 1985a			Lloyd et al. 1988
	++	*	1/1	Buffa et al. 1988b			
	++	*	1/1	Lloyd et al. 1988			
	–		0/5	Stefaneanu et al. 1988			
GH	+	*	9/9	DeStephano et al. 1984	ND		ND
	–	*	0/10	Lloyd et al. 1985a			
	++	*	2/2	Buffa et al. 1988b			
	+		2/5	Stefaneanu et al. 1988			
Prolactin	–	*	0/15	DeStephano et al. 1984	+	5/7	ND
	–	*	0/5	Lloyd et al. 1985a			Lloyd et al. 1988
	–	*	0/2	Buffa et al. 1988b			
	–	*	0/7	Lloyd et al. 1988			
	–		0/5	Stefaneanu et al. 1988			
TSH	+	*	3/7	Lloyd et al. 1985a	ND		ND
	++	*	1/1	Buffa et al. 1988b			
	–		0/5	Stefaneanu et al. 1988			
Null cell	++	*	7/7	DeStephano et al. 1984	+	5/6	ND
	+	*	12/17	Lloyd et al. 1985a			Lloyd et al. 1988
	++	*	2/2	Buffa et al. 1988b			
	++	*	6/6	Lloyd et al. 1988			
	+		2/9	Stefaneanu et al. 1988			
	++		1/1	Sobol et al. 1989			
Unspecified	+	*	6/10	Wilson & Lloyd 1984			
	+	*	5/6	Lloyd et al. 1986			
	++		9/11	Deftos et al. 1989			

Tab. 3-4 (continued)

Tumor	Chromogranin A	Chromogranin B	Secretogranin II
Prostate			
Carcinoid	- * 0/1 Turbat-Herrera et al. 1988	ND	ND
Small cell carcinoma	+ * 2/2 Turbat-Herrera et al. 1988	ND	ND
Mixed small cell adenocarcinoma	+ * 5/5 Turbat-Herrera et al. 1988	ND	ND
Skin			
Merkel cell tumor	++ * 2/3 Wilson & Lloyd 1984 ++ * 2/3 Buffa et al. 1988b ++* 1/1 Lloyd et al. 1988 + 1/1 Weiler et al. 1988	− 0/1 Lloyd et al. 1988 − 0/1 Weiler et al. 1988	ND
Thyroid			
C-cell carcinoma	++ * 5/5 Lloyd & Wilson 1983 ++ 2/2 O'Connor et al. 1983b ++ * 6/6 Wilson & Lloyd 1984 ++ *25/25 Sikri et al. 1985 ++ * 3/3 Varndell et al. 1985 ++ * 4/4 Lloyd et al. 1986 ++ 16/16 Schmid et al. 1987 ++ * 4/4 Lloyd et al. 1988 ++ *60/60 Schröder et al. 1988 ++ 1/1 Sobol et al. 1989	++ 16/16 Schmid et al.1987 ++ 4/4 Lloyd et al. 1988	+ 16/16 Schmid et al. 1987
calcitonin-negative			
Thymus			
Carcinoid	+ * 2/5 Herbst et al. 1987	ND	ND
NE carcinoma	+ * 1/1 Buffa et al. 1988b	ND	ND

Uterus

Endometrium

Argyrophilic carcinoma	+	*	19/36	Inoue et al. 1986	ND	ND
	–	*	0/3	Buffa et al. 1988b		
Small cell carcinoma	+	*	1/1	Manivel et al. 1986	ND	ND

Cervix

small cell carcinoma	+ *	5/10	Ulich et al. 1986	ND	ND
	+ *	11/13	Gersell et al. 1988		

Urinary bladder

small cell carcinoma	+ *	2/3	Blomjous et al. 1988	ND	ND

For each tissue, tumors are classified according to either histological appearance or hormone production, as referred to in the respective references.

++   all or the majority of tumor cells reported to be immunoreactive
+   the minority of tumor cells reported to be immunoreactive
−   no detectable immunoreactivity
ND   Not determined, NS   Not specified
*   Commercially available monoclonal antibody LK2H10 used to detect immunoreactive chromogranin A.
The numbers before and after the slash refer to the tumors found to be immunoreactive and the total tumors analyzed, respectively.

the lack of immunoreactivity for CgB and SgII in tumors of this category, recent preliminary data (Wiedenmann and Huttner, unpublished observations) suggest that at least some of these tumors may express CgB and SgII.

So far, only the expression of CgA has been determined in NE tumors of the breast, ovary, liver, middle ear, prostate, thymus, uterus, cervix and urinary bladder (see Tab. 3-4). With the exception of carcinoids, which appear to express CgA in a high percentage of the cases examined, all other NE tumors of these tissues examined showed only focal immunoreactions, and a relatively high number of the cases studied was immuno-negative.

CgA-positive NE tumors are not always positive for pancreastatin, whereas essentially all pancreastatin-positive tumors are also immunoreactive for CgA (Bishop et al. 1988; Schmidt et al. 1988a). This is to be expected since CgA, the precursor for pancreastatin, is not necessarily processed in every NE cell to immunoreactive pancreastatin.

In addition, the Cgs/Sgs may be useful tools in studying non-neoplastic human NE disease as well as diseases in which NE cells are, presumably, affected secondarily (e.g., Miller and Sumner 1982; Pietroletti et al. 1986; Rode et al. 1986).

3.3.3 Molecular Characterization of the Chromogranins/ Secretogranins in Human NE Tumors

For several types of human NE tumors displaying Cg/Sg reactivity in immunohistochemistry, the presence of these proteins has been confirmed by immunoblotting. For CgA, these tumors include pheochromocytomas, bronchial and intestinal carcinoids, islet cell tumors, parathyroid adenomas, small cell carcinomas, Merkel cell tumors, medullary thyroid carcinomas, paragangliomas and neuroblastomas (Lloyd and Wilson 1983; Schmid et al. 1987; Schober et al. 1987; Weiler et al. 1987, 1988; Lloyd et al. 1988; Wiedenmann et al. 1988). In most cases, a major immunoreactive polypeptide of M_r 75,000 identical to that noted in the adrenal medulla was observed.

For CgB, tumors found to be positive by immunoblotting include pheochromocytomas, bronchial and intestinal carcinoids, pituitary adenomas, medullary thyroid carcinomas and islet cell tumors (Schmid et al. 1987; Schober et al. 1987; Weiler et al. 1987, 1988; Lloyd et al. 1988; Wiedenmann et al. 1988). In most cases, immunoreactive polypeptides of M_r 110,000–120,000 and several immunoreactive bands of lower molecular weight were observed.

For SgII, tumors found to be positive by immunoblotting include pheochromocytomas, bronchial and intestinal carcinoids, islet cell tumors, medullary thyroid carcinomas, Merkel cell tumors and neuroblastomas (Schmid et al.

1987; Schober et al. 1987; Weiler et al. 1987, 1988; Wiedenmann et al. 1988). In most cases, immunolabeling revealed a reactive polypeptide of M_r 86,000, similar to SgII as observed in normal NE cells. In many of these cases, proteolytic fragments of the Cgs/Sgs were detected along with the full-length proteins, which were probably, at least in part, the results of processing events occurring *in vivo*.

3.3.4 Detection of Chromogranin/ Secretogranin mRNAs

The availability of cDNAs for the Cgs/Sgs also allows the detection of the corresponding mRNAs in normal and neoplastic human NE cells by *in situ* hybridization and/or Northern blot analyses. So far only few studies using this methodology have appeared (Deftos et al. 1986; Gazdar et al. 1988; Helman et al. 1988b; Siegel et al. 1988; Lloyd et al. 1989). In the case of the Cgs/Sgs, the detection of the mRNAs rather than the translation products may have several advantages. First, the detection of the mRNAs may in some instances be more reliable than that of the protein epitopes which, as outlined above, may not always be immunoreactive due to post-translational processing and epitope masking phenomena. Second, the cellular distribution of the mRNAs may differ from that of the corresponding proteins, although recent studies (Forss-Petter et al. 1989; Lloyd et al. 1989) on CgA and CgB did not reveal obvious discrepancies in this regard. However, considering a possible biological action of the Cgs/Sgs or their fragments on target cells, it cannot be excluded that some cells are immunoreactive for these proteins, not because they express them but because they endocytose them. Nerve growth factor provides a classical precedent for a striking difference in cellular distribution between the mRNA (which is found in the innervated non-neuronal cells) and the protein (which is accumulated in the innervating neurons; Davies et al. 1987). In such cases, the detection of the Cg/Sg mRNAs would allow a more reliable identification of NE cells than that of the respective proteins.

3.3.5 Detection of the Chromogranins/ Secretogranins in Human Plasma and Serum

The Cgs/Sgs and the peptides derived from them are physiologically released from the cells that produce them. In the case of CgA, the measurement of its concentration in the blood plasma has been shown to be useful in the diagnosis of NE tumors and the monitoring of their secretory activity (O'Connor and

Deftos 1986; O'Connor et al. 1989; Deftos et al. 1989; Moattari et al. 1989; Sobol et al. 1989). CgB, which is more abundant than CgA in human adrenal medulla and pheochromocytoma (Hagn et al. 1986; Benedum et al. 1987), has also been detected in human plasma and serum and appears to be present at higher concentrations than CgA (Wiedenmann et al. 1987). Since the expression of the individual Cgs/Sgs and their proteolytic processing in different NE cells is variable, the comparative determination of the concentration and pattern of these proteins and peptides in the plasma may well lead to a refined and easily applicable diagnostic procedure in NE disease (see chapter 7).

Acknowledgements

This review is an update of part of a recent review written by one of us (W.B.H.) together with Dr. Bertram Wiedenmann, University of Heidelberg Medical School. W.B.H. was supported by the Deutsche Forschungsgemeinschaft (SFB 317).

3.4 References

1 Ahn, T. G., Cohn, D. V., Gorr, S. U., Ornstein, D. L., Kashdan, M. A. and Levine, M. A. (1987). Primary structure of bovine pituitary secretory protein I (chromogranin A) deduced from the cDNA sequence. *Proc. Natl. Acad. Sci. USA* **84,** 5043–5047

2 Azzopardi, J. G., Evans, D. J. and Krausz, T. (1986). Endocrine differentiation in breast tumours. *Histopathology* **10,** 773–774

3 Bassetti, M., Huttner, W. B., Zanini, A. and Rosa, P. (1990). Colocalization of secretogranin/chromogranin with thyrotrophin and luteinizing hormone in secretory granules of cow anterior pituitary. *J. Histochem. Cytochem.* **38,** 1353–1363

4 Benedum, U. M., Baeuerle, P. A., Konecki, D. S., Frank, R., Powell, J., Mallet, J. and Huttner, W. B. (1986). The primary structure of bovine chromogranin A: a respresentative of a class of acidic secretory proteins common to a variety of peptidergic cells. *EMBO J.* **5,** 1495–1502

5 Benedum, U. M., Lamouroux, A., Konecki, D. S., Rosa, P., Hille, A., Baeuerle, P. A., Frank, R., Lottspeich, F., Mallet, J. and Huttner, W. B. (1987). The primary structure of human secretogranin I (chromogranin B): comparison with chromogranin A reveals homologous terminal domains and a large intervening variable region. *EMBO J.* **6,** 1203–1211

6 Benjannet, S., Leduc, R., Adrouche, N., Falgueyret, J. P., Marcinkiewicz, M., Seidah, N. G., Mbikay, M., Lazure, C. and Chretien, M. (1987). Chromogranin B (secretogranin I), a putative precursor of two novel pituitary peptides through processing at paired basic residues. *FEBS Lett.* **224,** 142–148

7 Bishop, A. E., Bretherton-Watt, D., Hamid, Q. A., Fahey, M., Shepherd, N., Valentino, K., Tatemoto, K., Ghatei, M. A., Bloom, S. R. and Polak, J. M. (1988). The occurence of pancreastatin in tumours of the diffuse neuroendocrine system. *Mol. Cell. Probes* **2**, 225–235

8 Bishop, A. E., Sekiya, K., Salahuddin, M. J., Carlei, F., Rindi, G., Fahey, M., Steel, J. H., Hedges, M., Domoto, T., Fischer-Colbrie, R. et al. (1989). The distribution of GAWK-like immunoreactivity in neuroendocrine cells of the human gut, pancreas, adrenal and pituitary glands and its colocalisation with chromogranin B. *Histochemistry* **90**, 475–483

9 Blaschko, H., Comline, R. S., Schneider, F. H., Silver, M. and Smith, A. D. (1967) Secretion of a chromaffin granule protein, chromogranin from the adrenal gland after splanchnic stimulation. *Nature* **215**, 58–59

10 Blomjous, C. E. M., Thunnissen, G. B. J. M., Vos, W., De Voogt, H. J. and Meijer, C. J. L. M. (1988) Small cell neuroendocrine carcinoma of the urinary bladder. *Virchows Arch. (A)* **413**, 505–512

11 Bordi, C., Pilato, F. P. and D'Adda, T. (1988) Comparative study of seven neuroendocrine markers in pancreatic endocrine tumours. *Virchows Arch. (A)* **413**, 387–398

12 Buffa, R., Mare, P., Gini, A. and Salvadore, M. (1988a) Chromogranins A and B and secretogranin II in hormonally identified endocrine cells of the gut and the pancreas. *Basic Appl. Histochem.* **32**, 471–484

13 Buffa, R., Rindi, G., Sessa, F., Fini, A., Capella, C., Jahn, R., Navone, F., De Camilli, P. and Solcia, E. (1988b) Synaptophysin immunoreactivity and small clear vesicles in neuroendocrine cells and related tumours. *Mol. Cell Probes* **1**, 367–381

14 Bussolati, G., Gugliotta, P., Sapino, A., Eusebi, V. and Lloyd, R. V. (1985) Chromogranin-reactive endocrine cells in argyrophilic carcinomas ("carcinoids") and normal tissue of the breast. *Am. J. Pathol.* **120**, 186–192

15 Bussolati, G., Papotti, M., Sapino, A., Gugliotta, P., Ghiringhello, B. and Azzopardi, J. G. (1987) Endocrine markers in argyrophilic carcinomas of the breast. *Am. J. Surg. Pathol.* **11**, 248–256

16 Chanat, E., Hubert, J.-F., Sion, B., DeMonti, M. and Duval, J. (1986) LHRH promotes the synthesis and release of a 87000 Da protein (GP-87) by enriched gonadotrophs. *Mol. Cell. Endocrinol.* **46**, 109–119

17 Chanat, E., Cozzi, M. G., Sion, B., de Monti, M., Zanini, A. and Duval, J. (1988) The gonadotrope polypeptide (GP 87) released from pituitary cells under luteinizing hormone-releasing hormone stimulation is a secretogranin II form. *Biochemie* **70**, 1361–1368

18 Chejfec, G., Falkmer, S., Grimelius, L., Jacobsson, B., Rodensjo, M., Wiedenmann, B., Franke, W. W., Lee, I. and Gould, V. E. (1987) Synaptophysin. A new marker for pancreatic neuroendocrine tumors. *Am. J. Surg. Pathol.* **11**, 241–247

19 Cohn, D. V., Elting, J. J., Frick, M. and Elde, R. (1984) Selective localization of the parathyroid secretory protein-I/adrenal medulla chromogranin A protein family in a wide variety of endocrine cells of the rat. *Endocrinology* **114**, 1963–1974

20 Cohn, D. V., Zangerle, R., Fischer-Colbrie, R., Chu, L. L. H., Elting, J. J., Hamilton, J. W. and Winkler, H. (1982) Similarity of secretory protein I from parathyroid gland to chromogranin A from adrenal medulla. *Proc. Natl. Acad. Sci. USA* **79**, 6056–6059

21 Cozzi, M. G., and Zanini, A. (1988) Secretogranin II is a Ca^{2+}-binding protein. *Cell. Biol. Int. Rep.* **12**, 493

22 Cozzi, M. G., Rosa, P., Greco, A., Hille, A., Huttner, W. B., Zanini, A. and De Camilli, P. (1989) Immunohistochemical localization of secretogranin II in the rat cerebellum. *Neuroscience* **28,** 423–441

23 Davies, A. M., Bandtlow, C., Heumann, R., Korsching, S., Rohrer, H. and Thoenen, H. (1987) Timing and site of nerve growth factor synthesis in developing skin in relation to innervation and expression of the receptor. *Nature* **326,** 353–358

24 Dayal,Y., Kirsten, A.,Tallberg, B. S., Nunnemacher, G., DeLellis, R. A. and Wolfe, H. J. (1986) Duodenal carcinoids in patients with and without neurofibromatosis. *Am. J. Surg. Pathol.* **10,** 348–357

25 Deftos, L. J., Murray, S. S., Burton, D. W., Parmer, R. J., O'Connor, D. T., Delageane, A. M. and Mellon, P. L. (1986) A cloned chromogranin A (CgA) cDNA detects a 2.3kb mRNA in diverse neuroendocrine tissues. *Biochem. Biophys. Res. Commun.* **137,** 418–423

26 Deftos, L. J., O'Connor, D. T., Wilson, C. B. and Fitzgerald, P. A. (1989) Human pituitary tumors secrete chromogranin-A. *J. Clin. Endocrinol. Metab.* **68,** 869–872

27 DeStephano, D. B., Lloyd, R. V., Pike, A. M. and Wilson, B. S. (1984) Pituitary adenomas. An immunohistochemical study of hormone production and chromogranin localization. *Am. J. Pathol.* **116,** 464–472

28 Ehrhart, M., Grube, D., Bader, M. F., Aunis, D. and Gratzl, M. (1986) Chromogranin A in the pancreatic islet: cellular and subcellular distribution. *J. Histochem. Cytochem.* **34,** 1673–1682

29 Ehrhart, M., Jörns, A., Grube, D. and Gratzl, M. (1988) Cellular distribution and amount of chromogranin A in bovine endocrine pancreas. *J. Histochem. Cytochem.* **36,** 467–472

30 Eiden, L. E. (1987) Is chromogranin a prohormone? *Nature* **325,** 301

31 Eiden, L. E., Huttner,W. B., Mallet, J., O'Connor, D. T.,Winkler, H. and Zanini, A. (1987) A nomenclature proposal for the chromogranin/secretogranin proteins. *Neuroscience* **21,** 1019–1021

32 Facer, P., Bishop, A. E., Lloyd, R. V.,Wilson, B. S., Hennessy, R. J. and Polak, J. M. (1985) Chromogranin: a newly recognized marker for endocrine cells of the human gastrointestinal tract. *Gastroenterology* **89,** 1366–1373

33 Falkensammer, G., Fischer-Colbrie, R., Richter, K. and Winkler, H. (1985) Cell-free and cellular synthesis of chromogranin A and B of bovine adrenal medulla. *Neuroscience* **14,** 735–746

34 Fasciotto, B. H., Gorr, S. U., DeFranco, D. J., Levine, M. A. and Cohn, D. V. (1989) Pancreastatin, a presumed product of chromogranin-A (secretory protein-I) processing, inhibits secretion from porcine parathyroid cells in culture. *Endocrinology* **125,** 1617–1622

35 Fischer-Colbrie, R., Schachiner, M., Zangerle, R. and Winkler, H. (1982) Dopamine beta-hydroxylase and other glycoproteins from the soluble content and the membranes of adrenal chromaffin granules. Isolation and carbohydrate analysis. *J. Neurochem.* **38,** 725–732

36 Fischer-Colbrie, R., Lassmann, H., Hagn, C. and Winkler, H. (1985) Immunological studies on the distribution of chromogranin A and B in endocrine and nervous tissues. *Neuroscience* **16,** 547–555

37 Fischer-Colbrie, R., Hagn, C. and Schober, M. (1987) Chromogranins A, B, and C: widespread constituents of secretory vesicles. *Ann. NY Acad. Sci.* **493,** 120–134

38 Fischer-Colbrie, R., Iacangelo, A. and Eiden, L. E. (1988) Neural and humoral factors separately regulate neuropeptide Y, enkephalin, and chromogranin A and B mRNA levels in rat adrenal medulla. *Proc. Natl. Acad. Aci.* USA **85,** 3240–3244

39 Fischer-Colbrie, R., Wohlfarter, T., Schmid, K. W., Grino, M. and Winkler, H. (1989) Dexamethasone induces an increased biosynthesis of chromogranin A in rat pituitary gland. *J. Endocrinol* **121,** 487–494

40 Forss-Petter, S., Danielson, P., Battenberg, E., Bloom, F. and Sutcliffe, G. J. (1989) Nucleotide sequence and cellular distribution of rat chromogranin B (secretogranin I) mRNA in the neuroendocrine system. *J. Mol. Neurosci.* **1,** 63–75

41 Funakoshi, S., Tamamura, H., Ohta, M., Yoshizawa, K., Funakoshi, A., Miyasaka, K., Tateishi, K., Tatemoto, K., Nakano, I., Yajima, H. et al. (1989) Isolation and characterization of a tumor-derived human pancreastatin-related protein. *Biochem. Biophys. Res. Commun.* **164,** 141–148

42 Gauweiler, B., Weihe, E., Hartschuh, W. and Yanaihara, N. (1988) Presence and coexistence of chromogranin A and multiple neuropeptides in Merkel cells of mammalian oral mucosa. *Neurosci. Lett.* **89,** 121–126

43 Gazdar, A. F., Helman, L. J., Israel, M. A., Russell, E. K., Linnoila, R. I., Mulshine, J. L., Schuller, H. M. and Park, J. G. (1988) Expression of neurendocrine cell markers L-dopa decarboxylase, chromogranin A, and dense core granules in human tumors of endocrine and nonendocrine orgin. *Cancer Res.* **48,** 4078–4082

44 Gerdes, H.-H., Phillips, E. and Huttner, W. B. (1988) The primary structure of rat secretogranin II deduced from a cDNA sequence. *Nucleic Acids Res.* **16,** 11811–11812

45 Gerdes, H.-H., Rosa, P., Phillips, E., Baeuerle, P. A., Frank, R., Argos, P. and Huttner, W. B. (1989) The primary structure of human secretogranin II, a widespread tyrosine-sulfated secretory granule protein that exhibits low pH- and calcium-induced aggregation. *J. Biol. Chem.* **264,** 12009–12015

46 Gersell, D., Mazonjian, F., Mutch, D. G. and Rindloff, M. A. (1988) Undifferentiated carcinoma of the cervix. A cliniopathologic ultrastructural and immunocytochemical study of 15 cases. *Am. J. Surg.* **12,** 684–699

47 Gorr, S.-U., Dean, W. L., Radley, T. L. and Cohn, D. V. (1988) Calcium-binding and aggregation properties of parathyroid secretory protein-I (chromogranin A). *Bone and Mineral* **4,** 17–25

48 Gorr, S. U., Shioi, J. and Cohn, D. V. (1989) Interaction of calcium with porcine adrenal chromogranin A (secretory protein-I) and chromogranin B (secretogranin I). *Am. J. Physiol.* **257,** 247–254

49 Greeley, G. H. Jr., Thompson, J. C., Ishizuka, J., Cooper, C. W., Levine, M. A., Gorr, S. U. and Cohn, D. V. (1989) Inhibition of glucose-stimulated insulin release in the perfused rat pancreas by parathyroid secretory protein-I (chromogranin-A). *Endocrinology* **124,** 1235–1238

50 Grino, M., Wohlfarter, T., Fischer-Colbrie, R. and Eiden, L. E. (1989) Chromogranin A messenger RNA expression in the rat anterior pituitary is permissively regulated by the adrenal gland. *Neuroendocrinology* **49,** 107–110

51 Grube, D., Aunis, D., Bader, F., Cetin, Y., Jorns, A. and Yoshie, S. (1986) Chromogranin A (CGA) in the gastroentero-pancreatic (GEP) endocrine system. I. CGA in the mammalian endocrine pancreas. *Histochemistry* **85,** 441–452

52 Hagn, C., Schmid, K. W., Fischer-Colbrie, R. and Winkler, H. (1986) Chromogranin A, B, and C in human adrenal medulla and endocrine tissues. *Lab. Invest.* **55,** 405–411

53 Hamid, Q. A., Bishop, A. E., Sikri, K. L.,Varndell, I. M., Bloom, S. R. and Polak, J. M. (1986) Immunocytochemical characterization of 10 pancreatic tumours, associated with the glucagonoma syndrome, using antibodies to separate regions of the pro-glucagon molecule and other neuroendocrine makers. *Histopathology* **10**, 119–133

54 Hartschuh, W. and Weihe, E. (1988) Multiple messenger candidates and marker substances in the mammalian Merkel cell – axon complex: a light and electron microscopic immunohistochemical study. *Progress Brain Res.* **74**, 181–187

55 Hartschuh,W.,Weihe, E. and Egner, U. (1989) Chromogranin A in the mammalian Merkel cell: cellular and subcellular distribution. *J. Invest. Dermatol.* **93**, 641–648

56 Hashimoto, S., Fumagalli, G., Zanini, A. and Meldolesi, J. (1987) Sorting of three secretory proteins to distinct secretory granules in acidophilic cells of cow anterior pituitary. *J. Cell Biol.* **105**, 1579–1586

57 Hearn, S. A. (1987) Electron microscopic localization of chromogranin A in osmium-fixed neuroendocrine cells with a protein A-gold technique. *J. Histochem. Cytochem.* **35**, 795–801

58 Helle, K. B. (1966) Some chemical and physical properties of the soluble protein fraction of bovine adrenal chromaffin granules. *Mol. Pharmacol.* **2**, 298–310

59 Helman, L. J., Ahn, T. G., Levine, M. A., Allison, A., Cohen, P. S., Cooper, P. S., Cooper, M. J., Cohn, D. V. and Israel, M. A. (1988a) Molecular cloning and primary structure of human chromogranin A (secretory protein I) cDNA. *J. Biol. Chem.* **263**, 11559–11563

60 Helman, L. J., Gazdar, A. F., Park, J. G., Cohen, P. S., Cotelingam, J. D. and Israel, M. A. (1988b) Chromogranin A expression in normal and malignant human tissues. *J. Clin. Invest.* **82**, 686–690

61 Herbst,W. M., Kummer,W., Hofmann,W., Otto, H. and Heym, C. (1987) Carcinoid tumors of the thymus. An immunohistochemical study. *Cancer* **60**, 2465–2470

62 Hogue-Angeletti, R. and Hickey,W. F. (1985) A neuroendocrine marker in tissues of the immune system. *Science* **230**, 89–90

63 Huttner, W. B. and Lee, R. W. H. (1982) Protein sulfation on tyrosine residues. *J. Cell Biol.* **95**, 389a

64 Huttner, W. B. and Benedum, U. M. (1987) Chromogranin A and pancreastatin. *Nature* **325**, 305

65 Huttner, W. B., Benedum, U. M. and Rosa, P. (1988) Biosynthesis, structure and functions of the secretogranins/chromogranins. In: *Molecular Mechanisms in Secretion.* (Thorn, N. A., Treiman, M. and Petersen, O. H. eds.). Munksgaard, Copenhagen, pp. 380–396

66 Hutton, J. C., Davidson, H. W., Grimaldi, K. A. and Peshavaria, M. (1987a) Biosynthesis of betagranin in pancreatic beta-cells. Identification of a chromogranin A-like precursor and its parallel processing with proinsulin. *Biochem. J.* **244**, 449–456

67 Hutton, J. C., Davidson, H. W. and Peshavaria, M. (1987b) Proteolytic processing of chromogranin A in purified insulin granules. Formation of a 20 kDa N-terminal fragment (betagranin) by the concerted action of a Ca^{2+}-dependent endopeptidase and carboxypeptidase H (EC 3.4.17.10). *Biochem. J.* **244**, 457–464

68 Hutton, J. C., Nielsen, E. and Kastern, W. (1988a) The molecular cloning of the chromogranin A-like precursor of beta-granin and pancreastatin from the endocrine pancreas. *FEBS Lett.* **236**, 269–274

69 Hutton, J. C., Peshavaria, M., Johnston, C. F., Ravazzola, M. and Orci, L. (1988b) Immunolocalization of Betagranin: A chromogranin A-related protein of the pancreatic B-cell. *Endocrinology* **122,** 1014–1020

70 Iacangelo, A, Affolter, H. U., Eiden, L. E., Herbert, E. and Grimes, M. (1986) Bovine chromogranin A sequence and distribution of its messenger RNA in endocrine tissues. *Nature* **323,** 82–86

71 Iacangelo, A., Okayama, H. and Eiden, L. E. (1988a) Primary structure of rat chromogranin A and distribution of its mRNA. *FEBS Lett.* **227,** 115–121

72 Iacangelo, A. L., Fischer-Colbrie, R., Koller, K. J., Brownstein, M. J. and Lee, E. E. (1988b) The sequence of porcine chromogranin A messenger RNA demonstrates chromogranin A can serve as the precursor for the biologically active hormone, pancreastatin. *Endocrinology* **122,** 2339–2342

73 Iguchi, H., Natori, S., Kato, K., Nawata, H. and Chretien, M. (1988) Different processing of chromogranin B into GAWK-immunoreactive fragments in the bovine adrenal medulla and pituitary gland. *Life Sci.* **43,** 1945–1952

74 Inoue, M., De Lellis, R. A. and Scully, R. E. (1986) Immunohistochemical demonstration of chromogranin in endometrial carcinomas with argyrophil cells. *Hum. Pathol.* **17,** 841–847

75 Johnson, T. I., Shapiro, B., Beierwaltes, W. H., Orringer, M. B., Lloyd, R. V. and Sisson, J. C. (1985) Cardiac paragangliomas. A clinicopathologic and immunohistochemical study of four cases. *Am. J. Surg. Pathol.* **9,** 827–834

76 Johnson, T. L., Zarbo, R. J., Lloyd, R. V. and Crissman, J. D. (1988) Paragangliomas of the head and neck: immunohistochemical neuroendocrine and intermediate filament typing. *Mod. Pathol.* **1,** 216–223

77 Kaufmann, E., Geisler, N. and Weber, K. (1984) SDS-PAGE strongly overestimates the molecular masses of the neurofilament proteins. *FEBS Lett.* **170,** 81–84

78 Kemper, B., Habener, J. F., Rich, A. and Potts, Jr. J. T. (1974) Parathyroid secretion: Discovery of a major calcium dependent protein. *Science* **184,** 167–169

79 Kiang, W.-L., Krusius, T., Finne, J., Margolis, R. U. and Margolis, R. K. (1982) Glycoproteins and proteoglycans of the chromaffin granule matrix. *J. Biol. Chem.* **257,** 1651–1659

80 Kimura, N., Sasano, N., Yamada, R. and Satoh, J. (1988) Immunohistochemical study of chromogranin in 100 cases of pheochromocytoma, carotid body tumour, medullary thyroid carcinoma and carcinoid tumour. *Virchows Arch. (A)* **413,** 33–38

81 Konecki, D. S., Benedum, U. M., Gerdes, H.-H. and Huttner, W. B. (1987) The primary structure of human chromogranin A and pancreastatin. *J. Biol. Chem.* **262,** 17026–17030

82 Krisch, K., Buxbaum, P., Horvat, G., Krisch, I., Neuhold, N., Ulrich, W., Srikanta, S. (1986) Monoclonal antibody HISL-19 as an immunocytochemical probe for neuroendocrine differentiation. Its application in diagnostic pathology. *Am. J. Pathol.* **123,** 100–108

83 Krisch, K., Horvat, G., Krisch, I., Wengler, G., Alibeik, H., Neuhold, N., Ulrich, W., Braun, O. and Hochmeister, M. (1988) Immunochemical characterization of a novel secretory protein (defined by monoclonal antibody HISL-19) of peptide hormone producing cells which is distinct from chromogranin A, B, and C. *Lab. Invest.* **58,** 411–420

84 Lassmann, H., Hagn, C., Fischer-Colbrie, R. and Winkler, H. (1986) Presence of chromogranin A, B and C in bovine endocrine and nervous tissues: a comparative immunohistochemical study. *Histochem. J.* **18,** 380–386

85 Lauweryns, J.M., van Ranst, K., Lloyd, R.V. and O'Connor, D.T. (1987) Chromogranin in bronchopulmonary neuroendocrine cells. Immunocytochemical detection in human, monkey, and pig respiratory mucosa. *J. Histochem. Cytochem.* **35**, 113–118

86 Lee, R.W.H. and Huttner, W.B. (1983) Tyrosine-O-sulfated proteins of PC12 pheochromocytoma cells and their sulfation by a tyrosylprotein sulfotransferase. *J. Biol. Chem.* **258**, 11326–11334

87 Linnoila, R.I., Mulshine, J.L., Steinberg, S.M., Funa, K., Matthews, M.J., Cotelingam, J.D. and Gazdar, A.F. (1988) Neuroendocrine differentiation in endocrine and nonendocrine lung carcinomas. *Am. J. Clin. Pathol.* **90**, 641–652

88 Lloyd, R.V. and Wilson, B.S. (1983) Sepcific endocrine tissue marker defined by a monoclonal antibody. *Science* **222**, 628–630

89 Lloyd, R.V., Blaivas, M. and Wilson, B.S. (1985b) Distribution of chromogranin and S100 protein in normal and abnormal adrenal medullary tissues. *Arch. Pathol. Lab. Med.* **109**, 633–635

90 Lloyd, R.V., Cano, M., Rosa, P., Hille, A. and Huttner, W.B. (1988) Distribution of chromogranin A and secretogranin I (chromogranin B) in neuroendocrine cells and tumors. *Am. J. Pathol.* **130**, 296–304

91 Lloyd, R.V., Iacangelo, A., Eiden, L.E., Cano, M., Jin, L. and Grimes, M. (1989) Chromogranin A and B messenger ribonucleic acids in pituitary and other normal and neoplastic human endocrine tissues. *Lab. Invest.* **60**, 548–556

92 Lloyd, R.V., Mervak, T., Schmidt, K., Warner, T.F. and Wilson, B.S. (1984) Immunohistochemical detection of chromogranin and neuron-specific enolase in pancreatic endocrine neoplasms. *Am. J. Surg. Pathol.* **8**, 607–614

93 Lloyd, R.V., Sisson, J.C., Shapiro, B. and Verhofstad, A.A. (1986) Immunohistochemical localization of epinephrine, norepinephrine, catecholamine-synthesizing enzymes, and chromogranin in neuroendocrine cells and tumors. *Am. J. Pathol.* **125**, 45–54

94 Lloyd, R.V., Wilson, B.S., Kovacs, K. and Ryan, N. (1985a) Immunohistochemical localization of chromogranin in human hypophyses and pituitary adenomas. *Arch. Pathol. Lab. Med.* **109**, 515–517

95 Manivel, C., Wick, M.R. and Sibley, R.K. (1986) Neuroendocrine differentiation in Müllerian neoplasms. An immunohistochemical study of a "pure" endometrial small-cell carcinoma and a mixed Müllerian tumor containing small-cell carcinoma. *Am. J. Clin. Pathol.* **86**, 438–443

96 Marcinkiewicz, M., Benjannet, S., Cantin, M., Seidah, N.G., and Chretien, M. (1986) CNS distribution of a novel pituitary protein '7B2': localization in secretory and synaptic vesicles. *Brain Res.* **380**, 349–356

97 Martens, J.M. (1988) Cloning and sequence analysis of human pituitary cDNA encoding the novel polypeptide 7B2. *FEBS Lett.* **234**, 160–164

98 McNutt, M.A. and Bolen, J.W. (1985) Adenomatous tumor of the middle ear. An ultrastructural and immunocytochemical study. *Am. J. Clin. Pathol.* **84**, 541–547

99 Miller, R.R. and Sumner, H.W. (1982) Argyrophilic cell hyperplasia and an atypical carcinoid tumor in chronic ulcerative colitis. *Cancer* **50**, 2920–2925

100 Miura, K. and Shirasawa, H. (1988) Primary carcinoid of the liver. *Am. J. Clin. Pathol.* **89**, 561–564

101 Moattari, A.R., Deftos, L.J. and Vinik, A.I. (1989) Effects of sandostatin on plasma chromogranin-A levels in neuroendocrine tumors. *J. Clin. Endocrinol. Metabl.* **69**, 902–905

102 Mooi, W. J., Dewar, A., Springall, D., Polak, J. M. and Addis, B. (1988) Non-small cell lung carcinomas with neuroendocrine features. A light microscopic, immuno-histochemical and ultrastructural study of 11 cases. *Histopathology* **13**, 329–337

103 Murphy, M. N., Glennon, P. G., Diocee, M. S., Wick, M. R. and Cavers, D. J. (1986) Nonsecretory parathyroid carcinoma of the mediastinum. Light microscopic, immunocytochemical, and ultrastructural features of a case, and review of the literature. *Cancer* **58**, 2468–2476

104 Murray, S. S., Burton, D. W. and Deftos, L. J. (1988) The coregulation of secretion and cytoplasmic ribonucleic acid of chromogranin-A and calcitonin by phorbol ester in cells that produce both substances. *Endocrinology* **122**, 495–499

105 Müller, J., Kirchner, T. and Müller-Hermelink, H. K. (1987) Gastric endocrine cell hyperplasia and carcinoid tumors in atrophic gastritis type A. *Am. J. Sur. Pathol.* **11**, 909–917

106 Nash, S. V. and Said, J. W. (1986) Gastroenteropancreatic neuroendocrine tumors. A histochemical and immunohistochemical study of epithelial (keratin proteins, carcinoembryonic antigen) and neuroendocrine (neuron-specific enolase, bombesin and chromogranin) markers in foregut, midgut, and hindgut tumors. *Am. J. Clin. Pathol.* **86**, 415–422

107 Navone, F., Jahn, R., Di Gioia, G., Stukenbrok, H., Greengard, P. and De Camilli, P. (1986) Protein 38: an integral membrane protein specific for small vesicles of neurons and neuroendocrine cells. *J. Cell Biol.* **103**, 2511–2527

108 Nesland, J. M., Holm, R. and Johannessen, J. V. (1986) A study of different markers for neuroendocrine differentiation in breast carcinomas. *Pathol. Res. Pract.* **181**, 524–530

109 Nolan, J. A., Trojanowski, J. Q. and Hogue-Angeletti, R. (1985) Neurons and neuroendocrine cells contain chromogranin: detection of the molecule in normal bovine tissues by immunochemical and immunohistochemical methods. *J. Histochem. Cytochem.* **33**, 791–798

110 O'Connor, D. T., Pandlan, M. R., Carlton, E., Cervenka, J. H. and Hslao, R. J. (1989) Rapid radioimmunoassay of circulating chromogranin A: *in vitro* stability, exploration of the neuroendocrine character of neoplasia, and assessment of the effects of organ failure. *Clin. Chem.* **35**, 1631–1637

111 O'Connor, D. T. and Bernstein, K. N. (1984) Radioimmunoassay of chromogranin A in plasma as a measure of exocytotic sympathoadrenal activity in normal subjects and patients with pheochromocytoma. *N. Engl. J. Med.* **311**, 764–770

112 O'Connor, D. T. and Deftos, L. J. (1986) Secretion of chromogranin A by peptide-producing endocrine neoplasms. *N. Engl. J. Med.* **314**, 1145–1151

113 O'Connor, D. T., Burton, D. and Deftos, L. J. (1983a) Chromogranin A: immunohistology reveals its universal occurrence in normal polypeptide hormone producing endocrine glands. *Life Sci.* **33**, 1657–1663

114 O'Connor, D. T., Burton, D. and Deftos, L. J. (1983b) Immunoreactive human chromogranin A in diverse polypeptide hormone producing human tumors and normal endocrine tissues. *J. Clin. Endocrinol. Metab.* **57**, 1084–1086

115 Oka, T., Yoshioka, T., Shrestha, G. R., Koide, T., Sonoda, T., Hosokawa, S., Onoe, K. and Sakurai, M. (1988) Immunohistochemical study of nodular hyperplastic thyroid glands in patient with secondary hyperparathyroidism. *Virchows Arch. (A)* **413**, 53–60

116 Ooi, A., Mai, M., Ogino, T., Ueda, H., Kitamura, T., Takahashi, Y., Kawahara, E. and Nakanishi, I. (1988) Endocrine differentiation of gastric adenocarcinoma. The

prevalence as evaluated by immunoreactive chromogranin A and its biological significance. *Cancer* **62,** 1096–1104

117 Orci, L., Ravazzola, M., Amherdt, M., Perrelet, A., Powell, S., Quinn, D. L. and Moore, H.-P. H. (1987) The trans-most cisternae of the Golgi complex: A compartment for sorting of secretory and plasma membrane proteins. *Cell* **51,** 1039–1051

118 Parmer, R. J., Koop, A. H., Handa, M. T. and O'Connor, D. T. (1989) Molecular cloning of chromogranin A from rat pheochromocytoma cells. *Hypertension* **14,** 435–444

119 Pelagi, M., Bisiani, C., Gini, A., Bonardi, M. A., Rosa, P., Mare, P., Viale, G., Grazia Cozzi, M., Salvadore, M., Zanini, A. et al. (1989) Preparation and characterization of anti-human chromogranin A and chromogranin B (secreto-granin I) monoclonal antibodies. *Mol. Cell Probes* **3,** 87–101

120 Piehl, M. R., Gould, V. E., Warren, W. H., Lee, I., Radoseich, J. A., Ma, Y. and Rosen, S. T. (1988) Immunohistochemical identification of exocrine and neuroen-docrine subsets of large cell lung carcinomas. *Path. Res. Pract.* **183,** 675–682

121 Pietroletti, R., Bishop, A. E., Carlei, F., Bonamico, M., Lloyd, R. V., Wilson, B. S., Ceccamea, A., Lezoche. E., Speranzy. V. and Polak, J. M. (1986) Gut endocrine cell population in coeliac disease estimated by immunocytochemistry using a monoclonal antibody to chromogranin. *Gut* **27,** 838–843

122 Rausch, D. M., Iacangelo, A. L. and Eiden, L. E. (1988) Glucocorticoid- and nerve growth factor-induced changes in chromogranin A expression define two different neuronal phenotypes in PC12 cells. *Mol. Endocrinol.* **2,** 921–927

123 Reiffen, F. U. and Gratzl, M. (1986) Ca^{2+} binding to chromaffin vesicle matrix proteins: effect of pH, Mg^{2+}, and ionic strength. *Biochemistry* **25,** 4402–4406

124 Rieker, S., Fischer-Colbrie, R., Eiden, L. and Winkler, H. (1988) Phylogenetic distribution of peptides related to chromogranins A and B. *J. Neurochem.* **50,** 1066–1073

125 Rindi, G., Buffa, R., Sessa, F., Tortora, O. and Solcia, E. (1986) Chromogranin A, B and C immunoreactivities of mammalian endocrine cells. Distribution, distinc-tion from costored hormones/prohormones and relationship with the argyrophil component of secretory granules. *Histochemistry* **85,** 19–28

126 Rode, J., Dhillon, A. P., Papadaki, L., Stockbrügger, R., Thompson, R. J., Moss, E. and Cotton, P. B. (1986) Pernicious anaemia and mucosal endocrine cell proliferation of the non-antral stomach. *Gut* **27,** 789–798

127 Rosa, P. and Zanini, A. (1981) Characterization of adenohypophysial polypeptides by two-dimensional gel electrophoresis. *Mol. Cell. Endocrin.* **24,** 181–193

128 Rosa, P. and Zanini, A. (1983) Purification of a sulfated secretory protein from the adenohypophysis. Immunochemical evidence that similar macromolecules are present in other glands. *Eur. J. Cell Biol.* **31,** 94–98

129 Rosa, P., Fumagalli, G., Zanini, A. and Huttner, W. B. (1985a) The major tyrosine-sulfated protein of the bovine anterior pituitary is a secretory protein present in gonadotrophs, thyrotrophs, mammotrophs and corticotrophs. *J. Cell Biol.* **100,** 928–937

130 Rosa, P., Hille, A., Lee, R. W., Zanini, A., De Camilli, P. and Huttner, W. B. (1985b) Secretogranins I and II: two tyrosine-sulfated secretory proteins common to a variety of cells secreting peptides by the regulated pathway. *J. Cell Biol.* **101,** 1999–2011

131 Rosa, P., Weiss, U., Pepperkok, R., Ansorge, W., Niehrs, C., Stelzer, E. H. and Huttner, W. B. (1989) An antibody against secretogranin I (chromogranin B) is packaged into secretory granules. *J. Cell Biol.* **109,** 17–34

132 Rundle, S., Somogyi, P., Fischer-Colbrie, R., Hagn, C., Winkler, H. and Chubb, I. W. (1986) Chromogranin A, B and C: immunohistochemical localization in ovine pituitary and the relationship with hormone-contaning cells. *Regulatory Peptides* **16,** 217–233

133 Said, J. W., Vimadalal, S., Nash, G., Shintaku, I. P., Heusser, R. C., Sassoon, A. F. and Lloyd, R. V. (1985) Immunoreactive neuron-specific enolase, bombesin, and chromogranin as markers for neuroendocrine lung tumors. *Hum. Pathol.* **16,** 236–240

134 Salahuddin, M. J., Sekiya, K., Ghatei, M. A. and Bloom, S. R. (1989) Regional distribution of chromogranin B 420–490-like immunoreactivity in the pituitary gland and central nervous system of man, guinea-pig and rat. *Neuroscience* **30,** 231–240

135 Scheithauer, B. W., Nora, F. E., Lechago, J., Wick, M. R., Crawford, B. G., Weiland, L. H. and Carney, J. A. (1986) Duodenal gangliocytic paraganglioma: A clinico-pathologic and immunocytochemical study of 11 cases. *Am. J. Clin. Pathol.* **86,** 559–565

136 Schmid, K. W., Fischer-Colbrie, R., Hagn, C., Jasani, B., Williams, E. D. and Winkler, H. (1987) Chromogranin A and B and secretogranin II in medullary carcinomas of the thyroid. *Am. J. Surg. Pathol.* **11,** 551–556

137 Schmidt, K. W., Weiler, R., Xu, R. W., Hogue-Angeletti, R., Fischer-Colbrie, R. and Winkler, H. (1989) An immunological study on chromogranin A and B in human endocrine and nervous tissues. *Histochem. J.* **21,** 365–373

138 Schmidt, W. E., Siegel, E. G., Kratzin, H. and Creutzfeldt, W. (1988b) Isolation and primary structure of tumor-derived peptides related to human pancreastatin and chromogranin A. *Proc. Natl. Acad. Sci.* USA **85,** 8231–8235

139 Schmidt, W. E., Siegel, E. G., Lamberts, R., Gallwitz, B., Creutzfeldt, W. (1988a) Pancreastatin: molecular and immunocytochemical characterization of a novel peptide in porcine and human tissues. *Endocrinology* **123,** 1395–1404

140 Schober, M., Fischer-Colbrie, R., Schmid, K. W., Bussolati, G., O'Connor, D. T. and Winkler, H. (1987) Comparison of chromogranins A, B and secretogranin II in human adrenal medulla and pheochromocytoma. *Lab. Invest.* **57,** 385–391

141 Schroeder, S., Boecker, W., Baish, H., Burk, G. G., Arps, H., Meiners, I., Kastendieck, H., Heitz, P. U. and Kloeppel, G. (1988) Prognosic factors in medullary thyroid carcinomas. Survivall in ralation to age, sex, stage, histology, immunocytochemistry and DNA content. *Cancer* **61,** 806–816

142 Seidah, N. G., Hendy, G. N., Hamelin, J., Paquin, J., Lazure, C., Metters, K. M., Rossier, J. and Chrétien, M. (1987) Chromogranin A can act as a reversible processing enzyme inhibitor. *FEBS Lett.* **211,** 144–150

143 Sekiya, K., Ghatei, M. A., Minamino, N., Bretherton-Watt, D., Matsuo, H. and Bloom, S. R. (1988) Isolation of human pancreastatin fragment containing the active sequence from a glucagonoma. *FEBS Lett.* **228,** 153–156

144 Siegel, R. E., Iacangelo, A., Park, J. and Eiden, L. E. (1988) Chromogranin A biosynthetic cell populations in bovine endocrine and neuronal tissues: detection by *in situ* hybridization histochemistry. *Mol. Endocrinol.* **2,** 368–674

145 Sietzen, M., Schober, M., Fischer-Colbrie, R., Scherman, D., Sperk, G. and Winkler, H. (1987) Rat adrenal medulla: levels of chromogranins, enkephalins, dopamine beta-hydroxylase and of the amine transporter are changed by nervous activity and hypophysectomy. *Neuroscience* **22,** 131–139

146 Sikri, K. L., Varndell, I. M., Hamid, Q. A., Wilson, B. S., Kameya, T., Ponder, B. A., Lloyd, R. V., Bloom, S. R. and Polak, J. M. (1985) Medullary carcinoma of

the thyroid. An immunocytochemical and histochemical study of 25 cases using eight separate markers. *Cancer* **56,** 2481–2491

147 Silver, M. M., Hearn, S. A., Lines, L. D. and Troster, M. (1988) Calcitonin and chromogranin A localization in medullary carcinoma of the thyroid by immuno-electron microscopy. *J. Histochem. Cytochem.* **36,** 1031–1036

148 Simon, J.-P., Bader, M.-F. and Aunis, D. (1988) Secretion from chromaffin cells is controlled by chromogranin A-derived peptides. *Proc. Natl. Acad. Sci.* USA **85,** 1712–1716

149 Simon, J. P., Bader, M. F. and Aunis, D. (1989) Effect of secretagogues on chromogranin A synthesis in bovine cultured chromaffin cells. *Biochem. J.* **260,** 915–922

150 Smith, A. D. and Winkler, H. (1967) Purification and properties of an acidic protein from chromaffin granules of bovine adrenal medulla. *Biochem. J.* **103,** 483–492

151 Sobol, R. E., Memoli, V. and Deftos, L. J. (1989) Hormone-negative, chromogranin A-positive endocrine tumors. *N. Engl. J. Med.* **320,** 444–447

152 Somogyi, P., Hodgson, A. J., DePotter, R. W., Fischer-Colbrie, R., Schober, M., Winkler, H. and Chubb, I. W. (1984) Chromogranin immunoreactivity in the central nervous system. *Brain Res.* **320,** 193–230

153 Stagno, P. A., Petras, R. E., Hart, W. R. (1987) Strumal carcinoids of the ovary. *Arch. Pathol. Lab. Med.* **111,** 440–446

154 Stefaneanu, L., Ryan, N. and Kovacs, K. (1988) Immunocytochemical localization of synaptophysin in human hypophyses and pituitary adenomas. *Arch. Pathol. Lab. Med.* **112,** 801–804

155 Steiner, H. J., Schmidt, K. W., Fischer-Colbrie, R., Sperk, G. and Winkler, H. (1989) Co-localization of chromogranin A and B, secretogranin II and neuropeptide Y in chromaffin granules of rat adrenal medulla studied by electron microscopic immunocytochemistry. *Histochemistry* **91,** 473–477

156 Tatemoto, K., Efendic, S., Mutt, V., Makk, G. and Feistner, G. J. (1986) Pancreastatin, a novel pancreatic peptide that inhibits insulin secretion. *Nature* **324,** 476–478

157 Tooze, J., Tooze, S. A. and Fuller, S. D. (1987) Sorting of progeny coronavirus from condensed secretory proteins at the exit from the trans-Golgi network of AtT20 cells. *J. Cell Biol.* **105,** 1215–1226

158 Tooze, S. A. and Huttner, W. B. (1990) Cell-free protein sorting to the regulated and constitutive secretory pathways. *Cell* **60,** 837–847

159 Tougard, C., Nasciutti, L. E., Picart, R., Tixier-Vidal, A. and Huttner, W. B. (1989) Subcellular distribution of secretogranins I and II in GH3 rat tumoral prolactin (PRL) cells as revealed by electron microscopic immunocytochemistry. *J. Histochem. Cytochem.* **37,** 1329–1336

160 Turbat-Herrera, E. A., Herrera, G. A., Gore, I., Lott, R. L., Grizzle, W. E. and Bonnin, J. M. (1988) Neuroendocrine differentiation in prostatic carcinomas. *Arch. Pathol. Lab. Med.* **112,** 1100–1105

161 Ulich, T. R., Liao, S. Y., Layfield, L., Romansky, S., Cheng, L. and Lewin, K. J. (1986) Endocrine and tumor differentiation markers in poorly differentiated small-cell carcinoids of the cervix and vagina. *Arch. Pathol. Lab. Med.* **110,** 1054–1057

162 Ulich, T. R., Kollin, M. and Lewin, K. J. (1988) Composite gastric carcinoma: report of a tumor of the carcinomacarcinoid spectrum. *Arch. Pathol. Lab. Med.* **112,** 91–93

163 Varndell, I. M., Lloyd, R. V., Wilson, B. S. and Polak, J. M. (1985) Ultrastructural localization of chromogranin: a potential marker for the electron microscopical recognition of endocrine cell secretory granules. *Histochem. J.* **17,** 981–992

164 Volknandt, W., Schober, M., Fischer-Colbrie, R., Zimmermann, H. and Winkler, H. (1987) Cholinergic nerve terminals in the rat diaphragm are chromogranin A immunoreactive. *Neurosci. Lett.* **81,** 241–244

165 Walts, A. E., Said, J. W., Shintaki, I. P. and Lloyd, R. V. (1985) Chromogranin as a marker of neuroendocrine cells in cytologic material – an immunocytochemical study. *Am. J. Clin. Pathol.* **84,** 273–277

166 Weiler, R., Feichtinger, H., Schmid, K. W., Fischer-Colbrie, R., Grimelius, L., Cedermark, B., Papotti, M., Bussolati, G. and Winkler, H. (1987) Chromogranin A and B and secretogranin II in bronchial and intestinal carcinoids. *Virchows Arch. (A)* **412,** 103–109

167 Weiler, R., Fischer-Colbrie, R., Schmid, K. W., Feichtinger, H., Bussolati, G., Grimelius, L., Krisch, K., Kerl, H., O'Connor, D. and Winkler, H. (1988) Immunological studies on the occurrence and properties of chromogranin A and B and secretogranin II in endocrine tumors. *Am. J. Surg. Pathol.* **12,** 877–884

168 Wick, M. R., Weatherby, R. P. and Weiland, L. H. (1987) Small cell neuroendocrine carcinoma of the colon and rectum: Clinical, histologic and ultrastructural study and immunohistochemical comparison with cloacogenic carcinoma. *Hum. Pathol.* **18,** 9–21

169 Wiedenmann, B., Waldherr, R., Rosa, P. and Huttner, W. B. (1987) Nachweis und Identifikation von Sekretogranin I im Gewebe und Serum von Patienten von neuroendokrinen Tumoren. *Z. Gastroenterologie* **8,** XXV, 589

170 Wiedenmann, B., Waldherr, R., Buhr, H., Hille, A., Rosa, P. and Huttner, W. B. (1988) Identifaction of gastroenteropancreatic neuroendocrine cells in normal and neoplastic human tissue with antibodies against synaptophysin, chromogranin A, secretogranin I (chromogranin B), and secretogranin II. *Gastroenterology* **95,** 1364–1374

171 Wilson, B. S. and Lloyd, R. V. (1984) Detection of chromogranin in neuroendocrine cells with a monoclonal antibody. *Am. J. Pathol.* **115,** 458–468

172 Winkler, H., Apps, D. K. and Fischer-Colbrie, R. (1986) The molecular function of adrenal chromaffin granules: established facts and unresolved topics. *Neuroscience* **18,** 261–290

173 Wohlfarter, T., Fischer-Colbrie, R., Hogue-Angeletti, R., Eiden, L. E. and Winkler, H. (1988) Processing of chromogranin A within chromaffin granules starts at C- and N-terminal cleavage sites. *FEBS Lett.* **231,** 67–70

174 Yoshie, S., Hagn, C., Ehrhart, M., Fischer-Colbrie, R., Grube, D., Winkler, H. and Gratzl, M. (1987) Immunological characterization of chromogranins A and B and secretogranin II in the bovine pancreatic islet. *Histochemistry* **87,** 99–106

175 Zanini, A. and Rosa, P. (1981) Characterization of adenohypophyseal polypeptides by two-dimensional gel electrophoresis. *Mol. Cell. Endocrinol.* **24,** 165–179

4 Neural Cell Adhesion Molecule NCAM in Neural and Endocrine Cells

Keith Langley and *Manfred Gratzl*

4.1 The Concept of Cell Type Markers

Considerable interest has centered over the last twenty years on the search for methods permitting the distinction between different cell types in complex tissues. While classical histological staining methods have proved invaluable in studying the cellular architecture of tissues, the search for more specific techniques that afford unambiguous identification of cell types was motivated by the fact that purely histological criteria frequently prove inadequate in other than normal adult tissues. In immature and in pathological tissues the microscopical appearance of cells is often radically different from that found in the healthy adult. In addition, increasing use of isolated cells grown in tissue culture has prompted the use of criteria other than those based on cellular morphology, which is rarely identical to that of the *in vivo* counterparts of cultivated cells and which may, moreover, be influenced by tissue culture conditions.

Although considerable progress towards this goal has been achieved by the development of histochemical methods, the major recent advances have followed the introduction and widespread application of immunocytochemical technology. This has dramatically altered both the way we recognize cells and indeed even the way in which we define a given cell type. Thus in the pre-immunocytochemical era cells were originally defined by their anatomical location and cellular morphology, which were correlated to their putative function. For example, silver impregnation methods have previously been very successfully employed to illustrate the delicate arborizations of large cerebellar neurons (Ramon y Cajal, 1905; Palay and Chan-Palay, 1974); in addition such staining methods suggested the function of such neurons in terms of their multiple synaptic connections with other cerebellar neurons. Current use of immunocytochemical methods performs the same task more easily and in a much more consistent manner (Mikoshiba et al., 1979; Reeber et al., 1981; Langley et al., 1982b; Lohmann et al., 1981; Roth et al., 1981). Similarly, astrocytes, which had early been visualized by Golgi methods as satellite cells with end-feet on blood vessels, may now be easily distinguished by the use of antibodies to the protein GFAP (**G**lial **F**ibrillary **A**cidic **P**rotein) considered to be specific for these cells (Dahl and Bignami, 1973; Eng and Bigbee, 1978). Such technology has led to a redefinition of astrocytes as cells which contain GFAP.

This specificity of expression of certain proteins by certain cell types forms the basis of current use of immunocytochemistry to identify cell types (for review see Langley et al., 1984). The precise mechanisms that lead to such specificity of expression of certain proteins are not yet clear. However it is evident that, while the genetic information contained in the cell nucleus after fertilization is potentially capable of coding for all proteins expressed by all cell types present in the adult organism, during embryogenesis regulatory mechanisms predispose the expression of certain of these proteins and determine the suppression of others. Pathological processes may nevertheless alter the normal course of events, resulting in up or down regulation of individual proteins in response to external stimuli. Moreover, cells in tissue culture have been shown to respond to external factors, including growth factors and hormones, by altered expression of "cell specific" proteins (e.g. Doherty and Walsh, 1987; Grant et al., 1988). However the mature normal cell may be defined as one in which the expression of a number of biochemical characteristics has reached a stable state. Thus, a given cell type which was previously defined only by morphological criteria may now be accorded a biochemical definition in terms of the range of proteins, glycoproteins and lipids expressed. While a large number of proteins are found to be present in many quite different cell types in the animal kingdom, a limited number may be expected to be associated with the specific nature or function of a given type and thus have a more restricted cellular distribution. This property of cell specificity of such molecules permits their use as marker substances to distinguish between different cell types. The value of these molecules as cell-type markers depends on their capacity to be expressed by a cell in both normal and abnormal (pathological or in culture) situations which may vary for certain proteins. This is particularly important when choosing a marker protein as a diagnostic tool in clinical practice. The elegance and precision of cell labeling methods have outweighed such minor disadvantages and have totally altered pathological laboratory practice for clinical diagnosis over recent years.

A second aspect of cell marker proteins should not be forgotten. It is reasonable to suppose that the exclusive or predominant expression of a given protein by a given cell type may be associated with the particular biological or physiological function of that cell. During the seventies the search for neuron specific proteins was prompted by the conviction that such proteins would be intimately associated with brain function. This is clearly the case for the enzymes involved in specific neurotransmitter metabolism or for the neuro-transmitters themselves (for review see Langley et al., 1984). Extensive neuropeptide mapping studies have indeed taught us a great deal about how the brain functions (Hökfelt et al., 1980). The link between cell marker protein and function is less evident in many other cases such as for example the S-100 protein (Matus and Mughal, 1975), the function of which remains a mystery long after the initial demonstration of its cellular specificity.

This chapter will concentrate principally on aspects of cellular expression but also briefly mention the likely function of what has been more recently revealed to be a family of related glycoproteins, collectively recognized under the name of neural cell adhesion molecule NCAM.

4.2 The NCAM Family

In order to better understand the cellular distribution of the different NCAM proteins, a brief historical account of their discovery and what is currently known of their biochemistry and molecular biology will be given. The discovery of this family of marker proteins was motivated by the search for molecules implicated in the construction of the extraordinary complexity of central nervous tissue. The adult human brain contains more than 10^8 neurons, each of which forms many synaptic connections. Such connections are not formed randomly but appear to follow overall design principles that lead to very specific inerconnecting patterns. The molecular mechanisms by which neurons select their target cells to produce such selective networks have intrigued neurobiologists for decades.

Three independent laboratories in two different continents applied different experimental approaches in different animal models to address this question. Over a relatively short period of time, each published reports on "brain specific molecules" which were subsequently found to have associated cell adhesive properties. Thus, Gerald Edelman's group in New York used an experimental paradigm, modified from that used in earlier searches for adhesion molecules, to select molecules involved in the reaggregation of dissociated chick retinal cells (for reviews see Edelman, 1983, 1984a,b; Rutishauser, 1984). He succeeded in isolating and purifying a molecule involved in such neural cell aggregation and referred to it as neural cell adhesion molecule. It was soon realized that this molecule was the chick equivalent of a molecule previously identified by crossed-electrophoresis by Elizabeth Bock's group in Copenhagen from rat brain, called D2 (Jørgensen and Bock, 1974; Jørgensen et al., 1980). This was one of several brainspecific proteins identified by this group at that time (Bock et al., 1975). Curiously, amongst these figured a molecule called synaptin, which was the first description of the vesicle marker protein now called synaptophysin (see chapter 2). A third laboratory, Christo Goridis' group in Marseille, using monoclonal antibody technology introduced in the 1970's obtained an antibody recognizing a **brain** cell **surface protein**, BSP-2, in the mouse (Hirn et al., 1981) which was subsequently shown to be the mouse equivalent of NCAM and D2 (Faissner et al., 1984; Noble et al., 1985).

The different data obtained on these three sets of molecules can thus now be collated to build up a more complete picture of the characteristics of what has been agreed since 1985 to be the same antigen, for which the name NCAM originally given by Edelman has been adopted. Since independant studies had been performed on what were originally considered to be unrelated molecules a number of initial contradictions appear when comparing the early literature on NCAM, D2 and BSP-2; for example whether or not glial cells express this molecule. While many of these differences no longer appear to exist certain species differences regarding the NCAM polypeptide forms expressed are evident; these will be discussed in more detail below. From the vast literature on NCAM several basic characteristics of this family can be singled out.

A unique gene coding for NCAM, containing nineteen principal exons, is present in the haploid genome. A mechanism of alternative splicing is currently considered (Rutishauser and Goridis, 1986; Owens et al., 1987; Goridis and Wille, 1988) to give rise to several species of mRNA the number of which varies according to species. In the chicken brain three NCAM mRNA species of 7.4, 6.7 and 4.3 have been detected and are thought to be translated into three principal NCAM polypeptides with molecular masses of approximately 180, 140 and 120 (referred to here as NCAM 180, NCAM 140 and NCAM 120, equivalent to NCAM **large d**omain NCAM **small d**omain and NCAM **small** surface **d**omain) (Cunningham et al., 1987). The two larger forms contain a single transmembrane region and differ essentially in the lengths of their cytoplasmic domains (Cunningham, 1986; Cunningham et al., 1987; Gennarini et al., 1984a,b; Nybroe et al., 1985; Barthels et al., 1987; Santoni et al., 1987), while the smaller isoform is anchored to the membrane by the intermediary of a phosphatidylinositol residue (He et al., 1987). The sequences external to the cell membrane of all three forms are essentially identical, containing five regions similar to those present in immunoglobulins (Barthels et al., 1987;

Cunningham et al., 1987; Williams, 1987), which has led to the classification of the NCAM gene as a member of the immunoglobulin super gene family. The homophilic binding site, in addition to a heparin binding site and a "hinge" region, are situated in the external domain of the molecule (Cole et al., 1986a,b; Cole and Glaser, 1986; Cole and Akeson, 1989; Frelinger and Rutishauser, 1986; Becker et al., 1989).

The situation appears to be rather more complex in rodents since in both the mouse (Gennarini et al., 1986; Barbas et al., 1988) and the rat (Small et al., 1987) five NCAM mRNA species are detectable of 7.4, 6.7, 5.2, 4.3 and 2.9 kb, two of which seem to encode for NCAM 120 (5.2 and 2.9 kb). These two mRNAs differ in the size of their noncoding regions, a first polyadenylation signal being used to generate the 2.9 kb mRNA and a second signal 2.3 kb downstream being used for the 5.2 kb message (Goridis and Wille, 1988; Barthels et al., 1987). By analogy with the chicken the 7.4 kb mRNA is considered to code for NCAM 180.

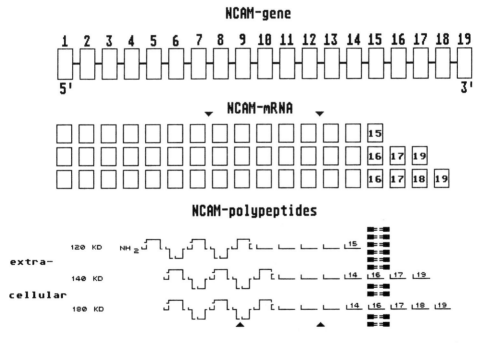

Fig. 4-1 Schematic representation of the NCAM gene, its transcription by alternative splicing and translation into the three principal NCAM polypeptides, illustrating the position of the sequences coded by different exons relative to the plasma membrane, the immunoglobulin-like domains and sites (arrowheads) of additional (tissue-specific) "minor" exons and their coded sequences.

Of the 19 major exons in the NCAM cDNA exons 1–14 are used to generate the external sequences of all three polypeptides, exon 15 codes for the anchoring sequence of NCAM 120 and exon 18 codes for the additional cytoplasmic 30–40 kD insert unique to NCAM 180 (Fig. 4.1 and Murray et al., 1986; Owens et al., 1987; Nybroe et al., 1988). Recently additional exons in the NCAM gene have been discovered, some of which code for large sequences in NCAM isoforms in certain tissues. Thus muscle NCAM mRNA contains an exon situated between exons 12 and 13 called msd (**muscle specific domain**) encoding a 37 amino acid sequence unique to this tissue (Walsh, 1988). Smaller additional exons have been found in rodent brain NCAM cDNA (Santoni et al., 1989). In particular exon π (30 bases) situated between exons 7 and 8 codes for a region in one of the immunoglobulin like domains, that may be close to the homophilic binding site and thus modify binding affinity (Santoni et al., 1989). Two more exons, one of only three base pairs, the other of 15, both situated between exons 12 and 13 code for a sequence in the hinge region of the molecule. These additional exons can give rise to minor heterogeneities in the three principal NCAM translation products, which may be fairly difficult to detect on immunoblots, since the molecular weights may differ very little from those of the principal forms. However S1 nuclease protection assays provide clearcut evidence of the existence in brain of multiple forms of mRNA coding for these additional polypeptides. Theoretically 24 different mRNA species are possible giving rise to 18 different NCAM polypeptides (Santoni et al., 1989).

The molecules are modified post-translationally by phosphorylation, sulphation and glycosylation (Lyles et al., 1984a; Nybroe et al., 1988). Glycosylation has been particularly well studied since major changes in levels of sialylation occur during normal brain development. Early so-called embryonic forms of NCAM contain high levels of polysialic acid groups, which markedly influences the adhesive binding properties of the molecule (Edelman, 1984; Rutishauser et al., 1985, 1988). NCAM isolated from adult tissue contains much less sialic acid. In addition to changes in polysialic acid, subsets of NCAM contain another carbohydrate hapten, HNK-1, consisting of 3'sulfated glucuronic acid (Noronha et al., 1986) recognized by antibodies first isolated as recognizing human natural killer cells (Schuller-Petrovic et al., 1983; Kruse et al., 1984). Thus, in summary, although three principal size classes of NCAM polypeptides have been found in both nervous and other tissues, all NCAM molecules of the same size class are not identical. Some may contain additional aminoacid sequences, others may be modified post-translationally in slightly different ways either containing different levels of polysialylation or different carbohydrate haptens (Williams et al., 1985).

4.3 Cellular Distribution of NCAM

4.3.1 Methodological Approaches

Two basic approaches have been employed to address the question of cellular NCAM expression; these may be broadly classified as morphological or biochemical.

4.3.1.1 Immunocytochemistry

The introduction by Coons et al. in 1941 of the use of flourescent probes covalently bound to antibodies and later modifications (Coons, 1961) provided the first opportunity of directly visualizing given protein tissue constituents at a cellular level. Considerable technical progress in immunocytochemical methods has subsequently eliminated many of the drawbacks of the early techniques and has led to the use of such methods as routine highly sensitive and specific staining procedures in pathological laboratories. Many techniques are currently available which differ in the probe or in the number of antibodies employed to visualize the molecule of interest. All depend for their success on the quality of the primary antibody i.e. its specificity and affinity for the antigen in question (Swaab et al., 1977). The higher the affinity of the antibody, the more it can be usefully diluted: the greater the dilution, the lower background non-specific staining will be, and thus the higher the specificity of the staining pattern.

This chapter will not attempt to review the advantages of individual methods. However it is important to remember that technical differences alone may give rise to apparently conflicting observations from different laboratories. In particular, the absence of staining may be simply due to the use of monoclonal antibodies with relatively lower affinities than commonly found with polyclonal antibodies. In addition monoclonal antibodies are sometimes more sensitive to tissue fixation procedures. Particular care should also be exercised with certain monoclonal antibodies that recognize cytoplasmic epitopes of NCAM, since cell permeabilization is a necessary prerequisite for their use.

Of the probes currently employed, fluorescent molecules such as fluoresceine isothiocyanate or rhodamine, or enzymes giving permanent preparations, such as peroxidase or alkaline phosphatase, have proved valuable for light microscopy. At the ultrastructural level peroxidase has proved to be the most versatile. Colloidal gold particles bound either to secondary antibodies or to protein A (Langley and Aunis, 1986; van den Pol et al., 1986) are particularly useful for labeling surface of antigens of isolated or cultured cells when morphometric analysis can be performed, but are limited by penetration problems when employed on pre-embedding tissue sections. The lack of penetration into ultrathin sections has limited the routine exploitation of post-embedding immunogold methods for NCAM localization.

4.3.1.2 Biochemical Approaches

Valuable information on cellular expression of individual NCAM glycoproteins has been obtained with combined immunochemical and protein separation techniques, which has only been partially possible with immunocytochemistry since antibodies specifically recognizing each of the three NCAM isoforms are not available. In particular turnover studies have been performed on NCAM isolated by immunoprecipitation followed by gel electrophoresis of extracts from cells labeled with radioactive precursors (Hirn et al., 1983; Lyles et al., 1984b; Nybroe et al., 1986; Linneman et al., 1985). More commonly electrophoresis and immunostaining of Western blots proves to be simpler and adequate for analyzing different NCAM isoforms in different tissues (Rougon et al., 1982; Hoffman et al., 1986; Nagata and Schachner, 1986). Quantitation is possible when radioiodinated second antibodies or protein A are used to locate the protein bands. The technique of crossed-electrophoresis (Bock, 1972) and ELISA have also provided quantitative estimates of different NCAM isoforms (Bock et al., 1983). These methods have been usefully applied in biosynthetic studies of NCAM in cultured neurons, glial cells and tumor lines. *In vivo* studies on NCAM extracted from the lateral geniculate nucleus after injection of radioactive precursors into the eye (which is followed by axonal transport of newly synthesized glycoproteins along the optic nerve) has afforded comparisons with the *in vitro* situation (Nybroe et al., 1986). Cell free translation of NCAM has also been investigated using crude RNA preparations from cultured cells and tissue homogenates (Nybroe et al., 1986).

Post-translational modifications have also been studied by immunochemical techniques. The most important of these as already mentioned concerns polysialic acid (PSA) content in the so-called embryonic form and is of special interest in pathology where frequently these forms reappear. Thus after nerve injury newly synthesized NCAM is found to be highly sialylated which is probably related to the need for less stability in intercellular contacts during the period of cellular redistribution and tissue repair (Daniloff et al., 1986a). The presence of high PSA forms of NCAM can be easily suspected from Western immunoblots by their appearance as broad diffuse bands of higher apparent Mr (200–250 kD). Pretreatment of such forms of NCAM with neuraminidase, which removes successive sialic acid residues from PSA chains, alters their anomalous electrophoretic profile from a broad smear to the descrete three banded pattern of adult brain NCAM (Rutishauser, 1984). This suggests that the three principle NCAM isoforms are present during embryonic life but are polysialylated.

Other post-translational modifications affecting the external domain of NCAM have been studied. Sub-populations of NCAM polypeptides may be distinguished by their complement of carbohydrate epitopes. The HNK-1 carbohydrate present on human natural killer cells (Abo and Balch, 1981) has been detected by successive immunoprecipitation with different antibodies on various adhesion molecules including L1, MAG and NCAM (Kruse et al., 1984; 1985; Poltorak et al., 1987). Only about 30 % of NCAM isolated from rodent brain carry this epitope. It has been suggested that this carbohydrate may itself be a ligand in cell adhesion (Künemund et al., 1988).

4.3.1.3 Hybridization Techniques

Since the isolation of clones of total or partial NCAM cDNA and their subsequent sequencing it has become feasable to examine the expression of the NCAM message either at a cellular level by the technique of *in situ* hybridization (where a specific oligonucleotide probe or a probe derived from a partial cDNA is hybridized to tissue sections) or in tissue RNA extracts by either Northern blots or S1 nuclease protection assays. This is a particularly important development since such approaches not only confirm the capacity of a cell to synthesize NCAM, but also provide valuable information on its regulation at the genetic level. Such approaches are complementary to studies employing antibodies, since in one case the message is detected while in the other its translation product, the localization of which is not always identical (Prieto et al., 1989). Furthermore the technique of S1 nuclease protection analysis provides information on the sequence of the bases in the NCAM mRNA as well as the precise nature of the exons present (Barthels et al., 1987; Santoni et al., 1989). However, in cells or tissues where NCAM turnover is slow it may prove much more difficult to detect low levels of the mRNA, while significant levels of NCAM peptides pose no problems for visualization with antibodies.

Up to the present time few reports of *in situ* hybridization studies have been published but, in contrast, mRNA analytical studies of tissue extracts are numerous (Covault et al., 1986; Murray et al., 1986; Gennarini et al., 1986) and have provided very interesting data on developmental changes. Most early studies on NCAM distribution concentrated on embryonic development where the greatest changes in expression may be expected to occur in parallel with important tissue induction events. In this section we will concentrate mainly on NCAM expression in different cell types of adult tissues, where the phenotype is stable, before discussing some of the early modifications, which are frequently transitory or recurrent during cellular migration or initial stages of tissue formation.

4.3.2 Neurons

4.3.2.1 Immunocytochemistry

One of the earliest references to a molecule that was subsequently found to be equivalent to NCAM was published in a report of a search for brain specific antigens (Jørgensen and Bock, 1974; Bock et al., 1974; 1975; Bock et al., 1980).

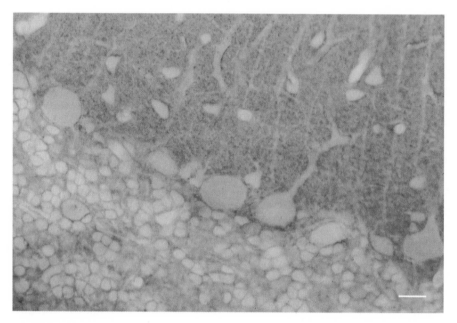

Fig. 4-2 Vibratome section of adult rat cerebellum immunoperoxidase labeled for NCAM.
Note the intense staining of the molecular layer, the unstained cytoplasm of Purkinje cells and marked staining of granule cells. Bar = 20 μm.

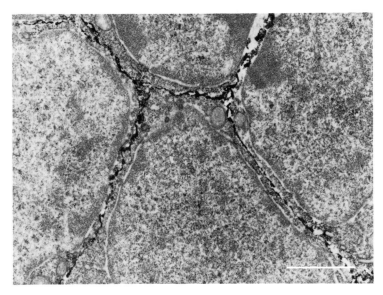

Fig. 4-3 Electron micrograph of rat cerebellar granule cells labeled for NCAM. The plasma membranes are intensely stained and cytoplasm is weakly immunoreactive. Section not counterstained. Bar = 1 μm.

The D2 antigen detected amongst several others, including protein 14–3-2 equivalent to neuron specific enolase was found to be absent from cultured astrocytes but enriched in synaptosomal fractions of rat brain. Though this antigen appeared not to be limited to nerve terminals a neuronal origin was concluded. At the same time, while much of the early work of Edelman's group concentrated on embryonic tissues, this group published direct immunocyto-chemical evidence of NCAM expression over the entire surface of mature sympathetic neurons, even though these were isolated from late embryonic chicks (Rutishauser et al., 1978a), with an antibody previously found to be capable of inhibiting embryonic retinal cell aggregation (Brackenbury et al., 1977; Rutishauser et al., 1978b). In contrast, a much more topographically restricted expression has been reported both in the retina (Jørgensen and Möller, 1981) and also in cerebellum where it was considered to be confined to synaptic regions of the neuronal membrane (Jørgensen and Möller, 1980). Later studies in both rodents and chickens provided convincing evidence of the widespread expression of NCAM by all types of neurons *in vivo* over their entire surface (see Figs. 4-2 and 4-3; Langley et al., 1982a, 1983; Daniloff et al., 1986b; Chuong et al., 1987). Such discrepancies are probably the result of differences in the affinity of antibodies used in different laboratories and also in different tissue fixation procedures. Immunoelectron microscopy did however reveal interesting differences between different types of cerebellar neurons (Langley et al., 1982a; 1983). NCAM was detected in granule cells, which constitute numerically the vast majority of cerebellar neurons, intracellularly in the perikaryon, although little was detected in their axons, in addition to at their cell surface. An intracellular localization was never observed for the large cerebellar neurons such as Purkinje cells, presumably because cytoplasmic levels were below limits of detection. These data were interpreted as reflecting differences of NCAM transport or turnover in different types of neurons. The question of turnover rates of NCAM in neurons of adult tissues is an intriguing one, which has been addressed in the biosynthetic studies of Elizabeth Bock's group, which reported that NCAM turnover was extremely slow in adult tissue. In fact it was calculated to decrease 350 fold during development (Linnemann et al., 1985). This would suggest that once NCAM is transported to the cell surface of Purkinje cells or to the distant parallel fibres (axons) of granule cells in normal adult cerebellum, it is subject to very slow degradation, resynthesis and reinsertion in the membrane. In contrast, synthesis of NCAM and transport through the Golgi apparatus to the cell membrane has been shown to be a rapid phenomenon in cultured neurons, lasting only several minutes (Lyles et al., 1984a). These data explain the absence or relatively low levels of NCAM observed immunocytochemically in the cytoplasm compared with the neuronal cell membrane. Though in healthy adult nervous tissue the NCAM synthetic machinery appears to be relatively silent, it is evident that external influences which may be provoked by traumatism can rapidly stimulate synthesis (Daniloff et al., 1986a). More recent immunocytochemical studies (Daniloff et al., 1986b) have confirmed that the adult NCAM staining pattern of chick nervous system reflects that of the late embryo (Thiery et al., 1982; Crossin et al., 1985) except for a diminution in intensity. This reduced NCAM immuno-labeling in adult tissue is particularly striking in myelinated fibre tracts such as spinal cord and optic nerve. Some reports suggest that the down-regulation of NCAM during development leads to its complete suppression in mature myelinated axons (Mirsky et al., 1986; Martini and Schachner, 1988). As already pointed out, caution in interpreting negative immunocytochemical data should be exercised since the detection of low antigen levels is much more dependant on technical factors. Nevertheless down regulation is confirmed by comparative analysis of both NCAM protein and NCAM mRNA levels in adults and young animals (Bock et al., 1980; Linnemann et al., 1985; Nybroe et al., 1986; Gennarini et al., 1986). It is likely that it is still present even on axons of CNS and PNS neurons in the adult though at much reduced levels and may also be associated with a slower turnover. In addition the reduced expression of NCAM in adult peripheral nerves seems also to be associated with changes in its topographical distribution on the axon surface: the membrane at nodes of Ranvier appear to contain more NCAM than is detectable on the internodal axon (Rieger et al., 1986). These highly specialized regions which predispose the rapid saltatory conduction of impulses in myelinated nerves were shown several years previously (Langley, 1979) to be the sites of high concentrations of acidic carboxylated polyanions (which could in part include sialic acid containing glycoproteins such as NCAM). Histochemistry with ruthenium red (a former botanical stain for pectin, a carboxyl acid rich polyanion) results in staining not only of the nodal axon membrane, but also the membrane of Schwann cells, which have subsequently been shown to contain NCAM (see below).

NCAM has also been shown to be present intraaxonally at nodes of Ranvier (Rieger et al., 1986), though the function associated with such a distribution at present remains obscure. Thus, in nervous tissues, a degree of compartmentalisation of NCAM expression may occur *in vivo* after permanent stable intercellular relationships are established. In contrast, studies on neurons *in vitro* (Nègre-Aminou et al., 1988) show no differences in total NCAM membrane density between cell bodies and neurite like extensions. This parallels the situation *in vivo* for non-myelinated fibres and may reflect a lesser degree of differentiation.

4.3.2.2 NCAM Isoforms in Neurons

In surveying the literature on the nature of NCAM polypeptides expressed by neurons the reader is confronted by apparent contradictions. Many studies have been performed on neuronal cultures and it is often difficult to extrapolate from cultures to adult neurons for several reasons. Cell isolation and culture conditions may be expected to modify NCAM synthesis since this has been shown to be sensitive to traumatism influencing cell contact relationships *in vivo*. Trophic factors may at least be partly responsible for such changes (Prentice et al., 1987). In addition most *in vitro* studies have employed neurons from embryonic tissues in which the stable adult NCAM phenotype has not been attained.

It is important first to consider species differences in the relative expression of individual NCAM polypeptides. Birds appear to express all three NCAM forms but their relative amounts and developmental pattern are different from those of rodents. The small surface domain peptide NCAM 120 is very weakly expressed in embryonic chick brain (Cunningham, 1986) compared to that of rodents and then only in late development. Contradictory data on frog NCAM has been published. In one report (Sunshine et al., 1987) only NCAM 180 was detected while Levi et al. (1987) reported that NCAM 140 was the first detectable form early during embryogenesis followed by NCAM 180. The appearance of NCAM 120 was shown to be a relatively late developmental event as in the chicken. Frog is unusual however in the expression by liver of an NCAM of 160 kD different to the brain isoforms (Levi et al., 1987).

Analysis of tissue extracts from different anatomical regions of the CNS at different developmental times have led to conclusions on cell type expression of different NCAM isoforms (Rougon et al., 1982; Chuong and Edelman, 1984). Different brain regions reflect both developmental biosynthetic trends in different compartments of a given cell type as well as the cellular composition of the tissue at a given time. Since glial cells also contain NCAM, though in lower amounts, tissue extracts correspond to a mixture of both neuronal and glial components. During CNS development neuronal proliferation and migration in general preceed glial cell multiplication and maturation. Thus total tissue NCAM at later stages will reflect a relatively greater contribution from glial cells compared to neurons. It has been found in general that NCAM 120 levels increase later than those of the other principal isoforms in developing brain. Analysis of adult spinal cord white matter demonstrates clearly that anatomical regional differences can give strong clues concerning cell type NCAM expression. Spinal cord contains relatively low amounts of NCAM 180 and 140 and larger amounts of NCAM 120 (Chuong and Edelman, 1984). Its NCAM composition may provide a better indication of astrocyte NCAM isoforms than neuronal forms, since this tissue is composed mainly of axons and glial cells and both axonal and oligodendrocyte NCAM is repressed when fibres are myelinated. Both sets of data suggest that both NCAM 180 and 140 are more typically neuronal than NCAM 120. The use of antibodies that specifically recognize NCAM 180 in immunocytochemical studies of the cerebellum indicates that this polypeptide is characteristic of post-mitotic neurons (Pollerberg et al., 1985) and suggests that this polypeptide is topographically compartmentalized. Higher concentrations of NCAM were found to exist at contact sites between cells and it may be enriched on cell extensions (Pollerberg et al., 1985; 1986). At the present time it would be unwise to generalize from these findings, even though in a quite different model, reinforcement of neuronal character which is accompanied by neurite outgrowth is also paralleled by higher NCAM 180 RNA levels and increased NCAM 180 synthesis (Doherty et al., 1988).

The capacity of mature neurons *in situ* to synthesize and axonally transport all three NCAM forms has been demonstrated for retinal ganglion neurons (Nybroe et al., 1986) and this agrees with data showing *in vitro* translation of all NCAM forms with cerebellar granule cells microsomes (i. e. post-mitotic neurons). Some reported immunochemical data of peripheral ganglia culture extracts also suggest that all three forms can be synthesized although NCAM 120 was minor, while other studies failed to detect NCAM 120 (Noble et al., 1985; Lyles et al., 1984a,b; Keilhauer et al., 1985; Nybroe et al., 1985). Such cultured cells may more closely resemble immature neurons.

4.3.2.3 "Embryonic" NCAM in Neurons

Recent immunocytochemical data have demonstrated the persistance of the carbohydrate polymer of "embryonic" NCAM with an antibody against the unusual highly sialylated NCAM (Rougon et al., 1986) in isolated regions of adult rat brain (Aaron and Chesselet, 1989; Rougon et al., 1990) in contrast to the widely held opinion that the embryonic-adult conversion is complete before adulthood. 30 % of adult frog CNS NCAM is in the highly sialylated form (Sunshine et al., 1987). Regions where high levels of PSA persist may be associated with the need for greater plasticity of neuronal contacts, which is afforded by the lower adhesive affinity of highly sialylated forms (Rutishauser et al., 1988). The relative immaturity of cultured neurons isolated from embryonic or young animals even when maintained in culture for prolonged periods is illustrated by the fact that the conversion to adult forms of NCAM is rarely observed (Nègre-Aminou et al., 1988), emphasizing the danger of extrapolating from cultures to adult tissues. Sialylation has been studied in acellular systems using lysates obtained from young and adult rats: sialylation of both NCAM 180 and 120 was observed with cerebellar granule cells, but the NCAM 140 isoform did not appear to be sialylated *in vitro* (Breen et al., 1987, see also Schlosshauer, 1989).

4.3.3 Astrocytes

The first report showing that astrocytes *in situ* express NCAM were made with monoclonal antibodies against BSP-2, before it was realized that this antigen was equivalent to NCAM (Langley et al., 1982a; Hirn et al., 1983). While clearcut evidence of immunolabeling was obtained on cerebellar tissue sections, curiously the same antibodies failed to immunolabel cultured astrocytes. This may have been due to the relative immaturity of cultured astrocytes compared with their *in situ* counterparts, the relatively low levels of astrocyte NCAM compared to neurons and the relatively low affinity of the monoclonal antibodies employed at that time. More recent immunocytochemical data with polyclonal antibodies (Noble et al., 1985) showed that cultured astrocytes indeed contain NCAM but quantitative analysis of immunolabeled cultured astrocytes (van den Pol et al., 1986) showed that membrane density was 25 fold less than that found on cultured (immature) neurons. Nevertheless, considerable variation in NCAM density was noted in neurons of the same culture. Immunoblot analysis of cultured astrocytes (Nybroe et al., 1986; Noble et al., 1985)showed a predominance of NCAM 140 with lower levels of NCAM

120, with little or no detectable NCAM 180. As suggested above for neurons, cell type characteristic NCAM isoforms can be deduced from total NCAM extracts from anatomical regions consisting mainly of one type of cell expressing NCAM. Optic nerve and spinal cord myelinated tracts contain no nerve cell bodies, axon levels of NCAM are relatively low in the adult and oligodendrocytes *in vivo* do not appear to express significant quantities of NCAM. Thus the relatively high levels of NCAM 120 and low levels of NCAM 180 found in adult rat spinal cord and optic nerve (our unpublished data) may be assumed to be mainly due to astrocytes. This agrees with the data from cultured astrocytes, but it may be argued that astrocytes in culture resemble "reactive" astrocytes.

Further indications that NCAM 120 is a predominant NCAM translation product in astrocytes are obtained from studies of mixed neuronal and astrocyte cultures of Friedlander et al. (1985) in which NCAM 120 synthesis was found to be preferentially inhibited compared to other isoforms by inhibitors of cell proliferation which affect mainly astrocytes in these cultures.

Pituicytes are the glial cells present in the neurohypophysis. While some morphological differences are notable between these cells and brain astrocytes, they resemble the latter with regard to the presence of GFA protein. Immunoelectron microscopy demonstrates that these cells, like brain astrocytes, also express NCAM (Fig. 4-5 and Langley et al., 1989a). Neurohypophysis has been found to contain little NCAM 120 (Langley et al., 1989a); it is possible that all is contributed by the pituicytes or that these cells have a different NCAM polypeptide phenotype than brain astrocytes.

Section of peripheral nerve (Daniloff et al., 1986a) or optic nerve (our unpublished data) produce a marked up-regulation of NCAM synthesis which in peripheral nerve is associated with a return to immature forms of NCAM. In optic nerve the massive astrogliosis associated with axonal degeneration is accompanied by a considerable increase in immunostaining intensity with anti-NCAM antibodies.

4.3.4 Oligodendrocytes

In contrast to astrocytes, no direct immunocytochemical evidence *in situ* has been obtained of NCAM in oligodendrocytes, the other major brain glial cell type. Nevertheless the capacity of cultured oligodendrocytes to synthesize NCAM has been unambiguously demonstrated (Trotter et al., 1989; Bhat and Silberberg, 1985, 1988), which once more emphasizes the danger of extrapolating from cultures to brain tissue. The profile of NCAM polypeptides expressed by cultured oligodendrocytes has been found to depend on their degree of maturation. Oligodendrocyte maturation has been followed both *in vivo* and *in vitro* by the successive appearance of the glycolipid antigens 04 and 01 (Trotter et al., 1989). Using these indices it was shown that immature oligodendrocytes express both NCAM 140 and 120 while more mature cells express predominantly NCAM 120. In addition the NCAM in these cultures is highly polysialated demonstrating the relative immaturity of these cells. *In vivo* it is likely that the reduction of NCAM expression noted in culture continues further until it is almost completely or totally repressed by the time myelination occurs.

4.3.5 Schwann Cells

Even though Schwann cells are the peripheral nervous system myelinating counterparts of oligodendrocytes, marked differences exist between them. Oligodendrocytes can myelinate many axons, while the Schwann cell can myelinate a single internode. Schwann cells are surrounded by a basement membrane *in situ* and their embryonic origin is the neural crest, while oligodendrocytes like astrocytes are derived from neural tube precursor cells. Curiously, Schwann cells express proteins such as S-100 which in the CNS has been employed as an astrocyte marker. Schwann cells like both oligodendrocytes and astrocytes have been shown to express NCAM in culture, but like the latter continue to express it *in vivo* in normal adult tissues. In addition, levels of NCAM in cultured Schwann cells are influenced by interactions with neurons (Seilheimer et al., 1989). *In vivo* immunolabeling is intense when Schwann cells surround unmyelinated fibres, but it has been reported that myelinating Schwann cells no longer contain NCAM (Mirsky et al., 1986; Covault and Sanes, 1986). It is likely that down regulation occurs but low levels appear to persist in peripheral nerves at nodes of Ranvier (Rieger et al., 1986). Interestingly, the HNK-1 epitope has also been localized at these sites in optic nerve (Ffrench-Constant et al., 1986). As is the case with other neural crest derivatives, the principal isoform present in Schwann cells is NCAM 140. In crushed or sectioned peripheral nerves NCAM synthesis dramatically increases and embryonic forms reappear (Daniloff et al., 1986a). Part of this increase is probably contributed by Schwann cells.

4.3.6 Endocrine Cells

The diffuse endocrine system encompasses a wide dispersity of either isolated cells or cell collectives situated in anatomically distant regions of the body. Cells of this system differ radically in terms of the nature of their secretory products and their physiological function, but their classification as a common series is due to data, accumulated over several years, demonstrating many shared features. These include their general secretory nature (endocrine not exocrine), certain morphological characteristics (in particular the presence of dense cored secretory granules visible in the electron microscope) and a number of biochemical parameters (see also chapter 1). Pearse (1968) was the originator of the concept of classifying neuroendocrine cells in a single series called APUD cells (amine precursor uptake and decarboxylation). An alternative terminology of neuroendocrine cells as paraneurons highlights similarities but also certain differences between them and neurons (Fujita, 1977). Certain endocrine cell types have been widely used as models for the study of secretory mechanisms, and as we have seen in chapter 2, parallels may exist with mechanisms regulating secretion at synapses in the central nervous system. Since the original APUD classification was proposed for these cells, data obtained with experimental approaches unavailable at that time have confirmed the common presence of "marker" proteins in different cell types of this series, which is one of the principal motivations behind this book. The multiplicity of common features of APUD cells led Pearse (1969) to suggest that all had a common embryonic (neural crest) origin. Subsequent advances in transplantation techniques have afforded the possibility of verifying this hypothesis. It has frequently been confirmed: chromaffin cells of the adrenal medulla may thus be viewed as modified neural crest derived sympathetic neurons in which certain neuronal characteristics such as neurite formation and neurofilament and synapsin expression are suppressed. In addition, the origin of the adenohypohysis, classically taught in medical schools to derive from the ectoderm of the stomodeum (primitive mouth), has been shown to have a neural origin, since cells destined to form this endocrine tissue are found in the neural ridge at early embryonic stages, but migrate away before formation of the neural crest (Couly and Le Douarin, 1985, 1987). Thus future hypophyseal cells may be subject to neural inductive influences before their arrival in the epithelial layer of the roof of the stomodeum. Notable exceptions to this hypothesis do however exist. The same elegant transplantation techniques demonstrating the neural origin of hypophysis have failed to confirm a similar origin for the pancreatic islet endocrine cells which are derived from cells situated in the ducts (see chapter 1; Le Douarin, 1978; Teitelman et al., 1987; Teitelman and Lee, 1987; Pearse, 1983).

4.3.6.1 Immunocytochemistry

The existence of certain common properties between neurons and neuroendocrine cells prompted us several years ago to investigate whether the same adhesive mechanisms between cells operate in brain and the endocrine system. The presence of NCAM was first tested in adrenal medullary cells with their confirmed neural crest origin (Langley and Aunis, 1984) and later extended to a wider range of endocrine tissues, including those of non-neural origin and tumor cells (Langley et al., 1987, 1989a).

The presence of NCAM in chromaffin cells of the adrenal medulla has been demonstrated or confirmed by light microscopy by several laboratories (Jørgensen and Richter-Landsberg, 1983) and biosynthetic studies show it to be

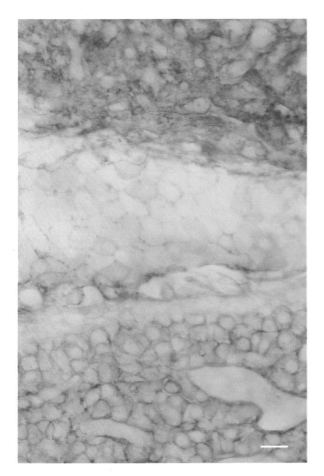

Fig. 4-4 Vibratome section of rat hypophysis immunoperoxidase labeled for NCAM.
Note the very intense staining of the neural lobe (top), the discrete surface labeling of cells in the intermediate lobe (middle) and the moderate surface labeling of anterior lobe cells (bottom). Bar = 20 μm.

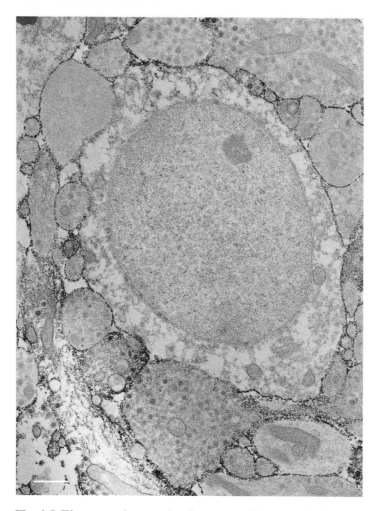

Fig. 4-5 Electron micrograph of rat neurohypophysis immunostained for NCAM, illustrating the surface staining of a pituicyte surrounded by neurosecretory cell, granule-containing, terminal swellings, which are also labeled on their surface. Section not counterstained. Bar = 1 µm.

synthesized in cultured rat chromaffin cells (Nybroe et al., 1986). Ultrastructural studies unequivocally show its membrane localization on both endocrine cell types (both adrenergic and noradrenergic) in this tissue. Little or no immunoreaction product has been detected intracellularly, which as for Purkinje cells in the cerebellum, may be the consequence of combined rapid transport and insertion into the plasmalemma after synthesis and a relatively slow turnover rate. Precise values on absolute amounts of NCAM in adrenal medulla are not available but in the adult rat, levels are much lower than in cerebellum (unpublished observations). Some of this is present on both unmyelinated nerve fibres and the Schwann cells surrounding them.

When other endocrine tissues were examined including the three lobes of the pituitary gland, the pancreas, tumor cell lines (Langley et al., 1989a) and certain human tumors (see below), an identical surface localization of NCAM was found immunocytochemically (Figs. 4-4–4-7). However, both qualitative and quantitative differences between different endocrine tissues have been found. For instance, in addition to the pronounced surface labeling of secretory cell processes, terminal swellings and nerve terminals constituting the neuroendocrine elements of the neurohypophysis, extracellular staining was observed using polyclonal anti-NCAM antibodies (Langley et al., 1989a). Extracellular localization of this molecule has also been reported in muscle (Sanes et al., 1986; Rieger et al., 1988). Since NCAM contains a heparin binding site (Cole and Glaser, 1986; Cole et al., 1986a,b) with a relative high affinity (Nybroe et al., 1989) capable of binding it to collagen fibrils by the intermediary of laminin, it is reasonable to suppose that any NCAM lost from the cell surface would adhere to neighbouring basement membrane components. Wether the loss of

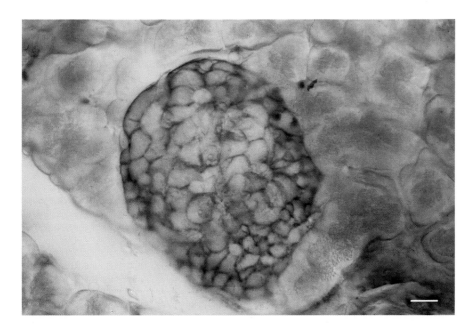

Fig. 4-6 Vibratome section of rat pancreas through an islet of Langerhans labeled for NCAM.
The surface labeling of all cells in the islet contrasts with the absence of staining in the exocrine pancreas. Bar = 20 μm.

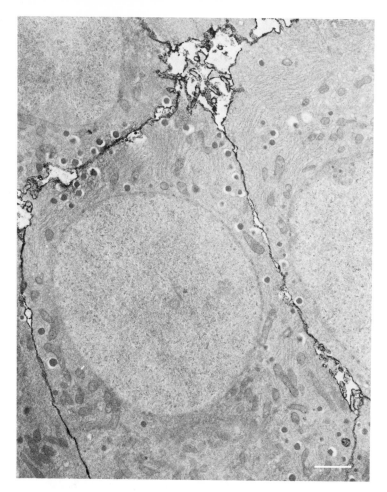

Fig. 4-7 Electron micrograph of rat endocrine pancreatic cells immunostained for NCAM.
The immunoreaction is confined to the plasma membranes. Section not counterstained.
Bar = 1 μm.

NCAM from cell surfaces is a normal phenomenon in healthy tissue is not known though a soluble secreted form of NCAM with a Mr 120 kD has been shown to exist (Bock et al., 1987). However, levels of NCAM 120 are very low in the neurohypophysis and at the present time the molecular nature of NCAM found extracellularly has not been resolved.

One of the striking features of the immunstaining pattern of rat pituitary gland sections concerns variations in labeling intensity found between the three lobes constituting this tissue (Fig. 4-4). The neural lobe is the most heavily stained and the anterior lobe is moderately labeled, while intermediate lobe endocrine cells appear very discretely labeled at their surface (Langley et al., 1989a). Apparent differences of intensity of immunolabeling should always be treated with caution in immunocytochemistry. Similar differences were however observed on immunoblots when extracts from identical amounts of adeno- and neurohypophysis were compared. It is, however, difficult to evaluate whether such overall differences in tissue concentrations reflect real differences in NCAM membrane densities in the two lobes, since their cellular composition differ radically: the anterior lobe of the hypophysis consists essentially of aggregates of cell bodies with few cell processes while the neurohypophysis, apart from a relatively small number of glial cells, which also express NCAM (Fig. 4-5), consists entirely of cell processes and terminals. Thus the membrane surface area is much larger on the basis of tissue weight in the latter compared to the former.

Pancreatic tissue is characterized by the absence of NCAM staining of exocrine cells and the marked surface labeling of all endocrine cells of the islets of Langerhans, irrespective of the nature of their secretory products (Figs. 4-6 and 4-7). Several years ago an NCAM related antigen was found to be present in rat testis localized on spermatids but not on mature spermatozoa (Jørgensen and Møller, 1983). Our recent preliminary studies on rat testis have also shown that groups of endocrine cells (Leydig cells) situated between the seminiferous tubules are immunoreactive with anti-NCAM sera. In addition, marked extracellular staining was observed in this tissue.

Molecular biological probes are currently being used to examine tissue distribution of the NCAM message. A major advantage of such techniques is their potential to study expression of individual exons in the transcript. Thus it is now possible to study not only the message for the major NCAM isoforms but also to detect the presence and analyze tissue distribution of the minor exons recently discovered. Using S1 nuclease protection assays the presence of NCAM mRNA has been confirmed in adrenal gland, neuro- and adenohypophysis and pancreas (our unpublished results).

4.3.6.2 NCAM Isoforms in Endocrine Tissues

Immunochemical analysis of NCAM polypeptide profiles of brain and endocrine tissue extracts revealed unexpected differences (Fig. 4-8 and Langley et al., 1987, 1989a). Although by immunocytochemistry with polyclonal anti-NCAM antibodies very similar cellular staining patterns are obtained with brain and endocrine tissues, the relative amounts of individual NCAM polypeptides in endocrine cells differ radically from those in brain: in addition one major exception was found amongst the endocrine tissues studied. NCAM

140 is the principal isoform expressed in rat adrenal gland, adenohypophysis and pancreatic islets but NCAM 180 is predominant in neurohypophysis. Cultured bovine adrenal medullary chromaffin cells may express measurable amounts of NCAM 180 (our own unpublished results; see also Nybroe et al., 1986) but this could be due to the increased tendency of such cells to express a neuronal phenotype, as illustrated by the appearance of neurofilament proteins in cultured chromaffin cells though these are undetectable *in vivo* (Grant et al., 1988). In addition, this could also be the result of species differences in the expression of NCAM 180 by endocrine cells. Few data exist at the moment on interspecies NCAM polypeptide expression in endocrine cells. Analysis of human tissues indicate that NCAM 140 is the predominant form present in adenohypophysis and endocrine lung tumors (Aletsee-Ufrecht et al., 1990b).

The neurohypophysis appears to represent an exception to the idea that the endocrine NCAM phenotype is NCAM 140. All three major NCAM polypeptides are found in this tissue although NCAM 120 is a minor constituent. The fact that the cell processes that form the bulk of this organ derive from perikarya situated in paraventricular and supraoptic nuclei of the hypothalamus, thus sharing similar embryonic origins with other central nervous system

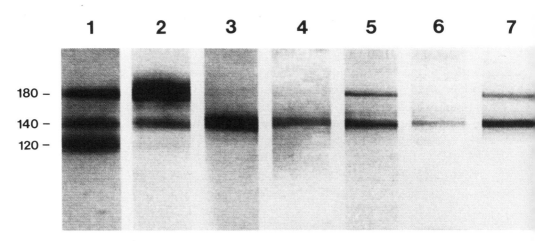

Fig. 4-8 Immunoblots showing NCAM isoforms in cerebellum (1), neurohypophysis (2), adenohypophysis (3), adrenal medulla (4), rat pheochromocytoma (5), pancreatic islets (6) and rat insulinoma cells (7).

Extracts of rat tissues and cells were resolved by polyacrylamide gel electrophoresis, transferred to nitrocellulose, and reacted with polyclonal anti-NCAM antibody. Immunoreactive bands were visualized with radiolabeled protein A and autoradiography. Molecular masses (kD) are indicated (from Aletsee-Ufrecht et al., 1988); by permission.

neurons, may be a sufficient reason to expect that their NCAM profile should more closely resemble that of CNS neurons. We have seen, however, that embryonic origin is not strictly determinant in NCAM expression. The structure of the neurohypophysis is quite different from that of other endocrine organs in that it is composed of cell processes that are very distant from their cell bodies. Studies on cultured neurones subjected to agents inducing neurite growth (Pollerberg et al., 1985, 1986) suggest that some compartmentalization of NCAM 180, perhaps due to an interaction with the cytoskeleton (Pollerberg et al., 1987), may occur between cell body and cell processes, the concentration being greater on the latter. Thus the higher levels of NCAM 180 found in neurohypophysis may be the result of such compartmentalization. In any case, these data show that all endocrine cells are not identical with regard to NCAM polypeptides, and that some may better suit the prefix "neuro-" in this context. It is interesting to note that the NCAM form that persists in adult endocrine cells is the form that is reported to be first expressed by very early embryos even before neural induction occurs (Levi et al., 1987; Jacobson and Rutishauser, 1986) and that NCAM 180 is considered by some to be the more characteristic form of differentiated post-mitotic neurons (Pollerberg et al., 1985, 1986). As is reviewed later, NCAM 140 is also the form found in certain strictly non-neural cells. Endocrine cells may thus be viewed as expressing more primitive less characteristically neuronal NCAM forms.

4.3.6.3 Endocrine Tumors

The PC12 cell line, derived from a rat adrenal medullary pheochromocytoma, has provided neurobiologists with a useful model for studying both endocrine cell biology and neuronal differentiation and maturation. Since these cells also express NCAM, their response to nerve growth factor NGF in neurite outgrowth formation has also provided the opportunity of studying NCAM expression during such phenomenon. Conflicting data have been published on the stimulatory effect of NGF (Doherty et al., 1988) or the absence of an effect (Friedlander et al., 1986) on certain forms of NCAM in PC12 cells. These differences may be the result of using different sublines of these cells. NCAM is essentially localized on the surface of PC12 cells as has been found for other endocrine and neural cells, and NCAM 140 is the predominant though not exclusive isoform present. Walsh's group observed both increased levels of NCAM 180 and NCAM mRNA of 7.2kb (the message considered to encode NCAM 180) after NGF treatment (Prentice et al., 1987), suggesting that this isoform may be more concentrated on neurite-like extensions. Elevated levels of this form have also been noted in another experimental endocrine tumor, a

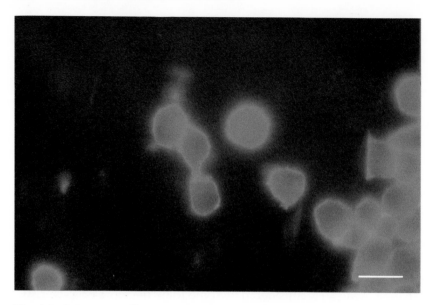

Fig. 4-9 Cultured rat insulinoma (RIN A2) cells labeled for NCAM by immunofluorescence. Note the pronounced surface labeling. Bar = 20 μm.

rat insulinoma (Fig. 4-9), compared with their normal counterparts (Langley et al., 1987). Rat insulinoma cells in culture tend to produce thin cell process unlike normal pancreatic β cells although these are less pronounced than those of NGF treated PC12 cells. Thus increased NCAM 180 expression may not be a directly related to the malignant transformation, but associated with cell morphological changes.

It is interesting to note that the expression of another CAM first identified in PC12 as NGF inducible large external (NILE) glycoprotein (McGuire et al., 1978) and subsequently shown to be equivalent to the cell adhesion molecule NgCAM of chick (Friedlander et al., 1986) and L1 in rodents (Bock et al., 1985) parallels that of NCAM 180 (Pigott and Kelly, 1986): it too is upregulated after NGF treatment. In fact the expression of L1 and NCAM 180 in many tissues is similar and may in part be due to an association between these two molecules (Thor et al., 1986) particularly on cell processes. Antibodies against L1 have been found to strongly stain neurohypophysis (Langley et al., 1989b;), which also contain high levels of NCAM 180: in contrast adenohypophysis and pancreatic islets are weakly or not at all stained for L1 nor do they contain NCAM 180. Chromaffin cells of the adrenal medulla represent an enigma, since only 20% appear to express L1 strongly in the rat adrenal which contains very little NCAM 180.

Two main classes of human endocrine tumors have been recently studied with regard to their NCAM expression; lung tumors and hypophyseal tumors. We have studied the small cell lung carcinoma (SCLC) on cell lines established from several patients (Fig. 4-10, Aletsee-Ufrecht et al., 1990a). These cell lines grow best in suspension culture but can be grown on coverslips, facilitating immunocytochemistry, where they retain a rounded cell morphology and their endocrine phenotype as exemplified by the presence of synaptophysin and neuron specific enolase. Immunoblot analysis has shown that NCAM 140 is the predominant if not exclusive NCAM form expressed. Other lung tumors examined included mesotheliomas, squamous cell carcinomas (i. e. non-endocrine tumors) and "large cell carcinoma". NCAM was found only in the latter: it is likely that these tumors are derived from SCLC. NCAM levels were found to vary between different SCLC patients, and even between two lines derived from tumors of the same patient at different times. Levels of other cell markers were also found to vary but in different directions. It can be concluded that the regulation of NCAM synthesis varies during the course of the disease and that such regulation parallels neither that of vesicle or souble proteins.

We have also recently examined human pituitary tumors for the presence of NCAM (Aletsee-Ufrecht et al., 1990b) including prolactinomas, growth hormone producing adenomas, and inactive adenomas, which were each characterized on the basis of clinical features, immunohistochemical data or

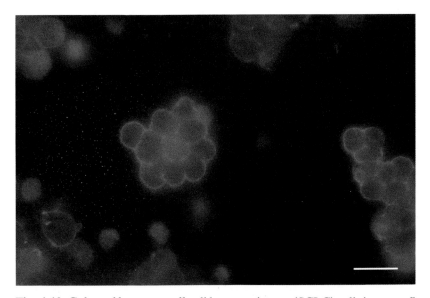

Fig. 4-10 Cultured human small cell lung carcinoma (SCLC) cells immunofluorescently labeled for NCAM showing marked surface labeling. Bar = 20 μm.

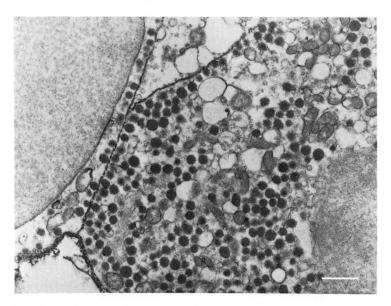

Fig. 4-11 Electron micrograph of human growth hormone producing adenoma immunostained for NCAM, illustrating the intense surface labeling of these cells. Section not counterstained. Bar = 1 μm (from Aletsee-Ufrecht et al., 1990b, by permission of Elsevier, Amsterdam).

serum hormone analyses. NCAM 140 was detectable on immunoblots of all but prolactinomas. Highly sensitive immunocytochemical methods confirmed a surface localization of NCAM in all these tumors (Fig. 4-11) even in prolactinomas, although staining intensity was much less in prolactinomas, confirming reduced levels in this tumor.

The presence of NCAM in certain human tumors may also be responsible for the staining observed with other anti-tumor antibodies of previously undetermined antigen specificity. From a number of monoclonal antibodies developed against a variety of tumors, some have been shown to recognize the NCAM molecule (Patel et al., 1989). In addition the staining of tumors with anti HNK-1 (Leu7) antibodies (chapter 7; Willison et al., 1986) could be due to a subset of NCAM molecules carrying the HNK-1 carbohydrate hapten.

4.3.7 Fibroblasts

Fibroblasts represent a further ambiguous class of cells on which apparently contradictory data have been published. Earlier reports produced no evidence of NCAM in fibroblasts. Fibroblasts in normal peripheral nerve in endo-, peri- and epineural situations are NCAM negative in adult tissue and also at all developmental ages studied (Seilheimer and Schachner, 1988; Martini and Schachner, 1988; Seilheimer et al., 1989). Fibroblasts isolated from skin and brain have not been found to contain NCAM. In contrast, immunocytochem- istry demonstrates NCAM expression by fibroblast-like cells which accumulate in nerve stumps after crush lesions or section of peripheral nerves and by perisynaptic fibroblasts in denervated muscle (Gatchalian et al., 1989). Fibroblasts isolated from both peripheral nerve and skeletal muscle and subsequently maintained in culture also have the capacity to express NCAM (Gatchalian et al., 1989). The isoform expressed is NCAM 140, which is the same as that found in peripheral nerve Schwann cells. Mesenchymal cells possibly related to fibroblasts in regenerating newt limbs have also been found to be NCAM positive (Maier et al., 1986). The capacity to express NCAM may thus be confined to particular subsets of fibroblasts or be induced either in culture or in pathological situations in which increased NCAM synthesis is known to occur: NGF activity also increases in peripheral nerves after injury.

4.3.8 Muscle

The considerable number of studies devoted to both adult and developing skeletal muscle since the initial discovery of NCAM in embryonic muscle (Grumet et al., 1982) have provided valuable insight into the regulation of NCAM synthesis and its molecular biology. We will not deal in detail here with muscle and the reader is referred to a recent review (Walsh, 1988) for a more detailed analysis. A brief discussion however of NCAM in muscle is pertinent in the context of its potential as a neural cell marker. It was the first tissue outside the nervous system in which NCAM expression was found to persist into adulthood. While NCAM is detectable in all primitive germ cell layers in early embryos, defined tissue inductive events lead either to its suppression or to its continued expression (Edelman, 1984). During skeletal muscle histogenesis, down-regulation of NCAM during myotube maturation and innervation is associated with a dramatic change in its localization on the surface membrane of the muscle cell. The wide distribution over the entire myotube surface is replaced in the adult by restricted expression over specialized contact regions, the neuromuscular junctions. This adult topographical distribution can be reversed to that found in immature muscle by denervation.

Analysis of the polypeptide nature of muscle NCAM illustrates how different tissues can process these molecules in specific ways. Initial immunoblot studies on muscle suggested a molecular mass for this antigen similar to that of brain NCAM 140. Further investigations showed that muscle NCAM mRNA contains an additional exon, not present in brain, encoding an amino acid sequence unique to muscle, and it was suggested that the isoform present in differentiated myotubes is equivalent to a modified NCAM 120, which, like this brain isoform, is anchored to the plasma membrane by phosphatidyl inositol (Moore et al., 1987).

4.3.9 Other Non-Neural Tissues

A rare but unusually malignant kidney tumor (Wilms tumor) represents a particularly interesting example of NCAM re-expression in pathological tissue. NCAM is detectable during early development of the kidney, but it is subsequently down-regulated and is absent from the adult gland (Roth et al., 1987). In Wilms tumour NCAM has been found to be not only strongly expressed, but as the highly sialylated "embryonic" form (Roth et al., 1988). The malignant transformation may thus be interpreted in terms of a reversion of the normal differentiation and maturation process leading to a cell with more primitive characteristics. Elevated levels of high PSA containing NCAM undoubtedly influences the contact relationships and "social behaviour" of these cells, properties which many years ago had been suggested to be linked to cell surface sialic acid content (Langley and Ambrose, 1964).

NCAM has been detectable in a few other non-neural tissues. Several adhesion molecules have been fond in rodent intestine with different distributions (Thor et al., 1986). NCAM appears to be confined to mesenchymal and neuroectodermally derived parts of the intestine. In particular it was found to be located in the outer muscle layers, not only in Auerbach's plexus but also in mesenchymal parts, and it was present in the submucosa, the muscularis mucosa, the lamina propria surrounding the crypts and in the inner part of villi. Surprisingly, all three NCAM isoforms were reported to be present, thus reflecting the CNS pattern rather than that found in peripheral nervous or non-nervous tissues. Individual NCAM containing cellular elements were unfortunately not identified in this study. Thus it is not known if intestinal endocrine cells contain NCAM. Some staining in the muscularis mucosa and in the lamina propria may be due to neurons or their processes.

Chicken skin has been found to contain NCAM (Chuong and Edelman, 1985a,b) which is similar or identical to that of brain except in its polypeptide profile: only NCAM 140 is detectable in skin (Murray et al., 1986). This persists into adulthood and is subject to the embryonic-adult post-translation modification described for the brain antigen. It is associated with the morphogenesis of chicken feathers, a process which continues throughout adult life. A cyclic pattern of both NCAM and LCAM (equivalent to the calcium dependent cell adhesion molecule E-cadherin) expression is found in growing and regenerating feather buds. In addition to membrane expression, NCAM is also located in certain extracellular regions after cell disappearance. NCAM mRNA has also been detected by the blot hybridization technique in embryonic heart, breast muscle, gizzard, lung and kidney (Murray et al., 1986). In non-neural tissues NCAM 180 mRNA has been only detectable in association with nervous elements such as the nerve plexuses in developing lung and the large enteric ganglia of gizzard. Most data suggest that when NCAM is present in non-neural tissues NCAM 140 is the isoform generally found.

4.4 Expression of NCAM During Development

Attempts to relate NCAM function to tissue structure and in particular to tissue inductive events have strongly motivated studies of its expression throughout embryonic development. Major "decisions" which determine the course of histogenesis are made relatively early in embryonic life. Radical changes in tissue levels of cell adhesion molecules which either stabilize intercellular contacts or promote cellular migration may be expected to coincide with this period of intense activity of cell sorting and reorganisation. The pattern of NCAM expression is indeed extremely plastic during early embryonic life (Edelman, 1983, 1984b). This is also true in adult tissues subject to inductive changes (Chuong and Edelman, 1985a,b). It is beyond the scope of this chapter to review in detail such complex changes in NCAM regulation during nervous tissue and endocrine tissue development. Readers are referred to the excellent studies by Edelman's group. In pioneering studies Thiery and coworkers (1982) reported the very early appearance of NCAM in embryonic tissues. NCAM is detectable in all three primitive germ layers, the endoderm, the mesoderm and the ectoderm, and has been shown to be present in those regions concerned with primary developmental axes (neural plate, neural tube, notochord, somites) and in those regions in which later inductive events occur such as the neural crest, optic, otic and pharyngeal placodes, cardiac mesoderm, mesonephric primordium and limb buds. When tissue inductive events occur, NCAM expression may be either suppressed permanently or temporarily or maintained according to the nature of induction.

A number of studies have addressed the problem of changing levels of NCAM during nervous tissue development. Levels vary between different brain regions or tissues during development (Bock et al., 1983; Linnemann et al., 1985). Levels are reduced more in myelinated tracts than in regions containing cell bodies (Daniloff et al., 1986b). Cerebellar development is a relatively late event compared with rat forebrain, occurring during the first three weeks of postnatal life. NCAM levels are low before birth but increase dramatically during the period of parallel fibre (granule cell axon) formation and synaptogenesis corresponding to a period of dramatic increase of neuronal surface area (Jørgensen and Honegger, 1983). Cerebellar NCAM subsequently decreases to a fairly constant level. This may be correlated with data from biosynthetic studies on cultures of rat brain cells obtained at embryonic day 17 and at postnatal day 25, which demonstrate a 50 % reduction in overall level (Nybroe et al., 1986). Analysis of NCAM mRNA levels show that the levels of the transcription product are reduced during brain development three times more than that of the translation products (Gennarini et al., 1986). This may be interpreted in terms of a slower NCAM turnover as development proceeds or a higher NCAM mRNA stabilization. A slower turnover rate of NCAM has been suggested after insertion into plasma membranes after axonal transport in the optic system (Garner et al., 1986). Interestingly, even within the cerebellum a differential distribution of individual NCAM polypeptides has been found during development (Pollerberg et al., 1985). While NCAM is detectable by immunocytochemistry with polyclonal antibodies recognizing all NCAM polypeptides at earliest times in the germinative layer of the cerebellum, NCAM 180 is not detectable in this zone. This polypeptide appears to be confined, in the molecular layer, to post-mitotic neurons. Regional differences have also been noted using specific probes capable of distinguishing between NCAM 180 and NCAM 140 messages (Murray et al., 1986). The NCAM 180 message has been detected as early as the third day of embryonic life in chick spinal cord but was not found in neuroepithelium. Although NCAM 140 message was detectable in embryonic heart, breast-muscle, skin, gizzard lung and kidney, NCAM 180 mRNA was found to be absent from these tissues and only detectable in tissues containing neural elements (such as the large enteric ganglia in gizzard or the nerve plexuses in oesophagus and lung buds). Very recent *in situ* hybridization studies confirm these earlier findings (Prieto et al., 1989).

A particularly relevant example of developmental NCAM up- and down-regulation in the context of this chapter is found in cells which constitute the adrenal medulla. The endocrine chromaffin cells are derived, as are sympathetic ganglia, from NCAM positive precursors in the neural crest (Thiery et al., 1982). When these cells migrate away from the neural crest towards their future site NCAM expression is suppressed. At the same time substrate adhesion molecules are upregulated (Duband et al., 1986). The reaggregation of these cells during the histogenesis of the adrenal gland coincides with renewed expression of NCAM 140, and is accompanied by down-regulation of fibronectin receptors. These data have been interpreted in terms of the stabilizing effect of NCAM on intercellular contacts, which would be counter-productive during the migration phase, but which are critical during formation of the adrenal medulla. In addition, since the cortical cells surrounding the centrally situated groups of adrenal medullary chromaffin cells do not express NCAM, a mechanism of "cell sorting" may operate in defining the particular tissue architecture of this endocrine gland (see Friedlander et al., 1989).

4.5 Post-Translational Developmental Modifications

The dramatic modifications in the expression of NCAM that occur during development or in post-traumatic or pathological situations underlined earlier can take a number of different forms. We have seen that tissue levels of total NCAM, and accordingly membrane density, is dependant on endogenous environmental stimuli. Such factors may also differentially influence the transcription of individual mRNA species coding for the different NCAM isoforms independently of each other. It is not yet known if these various NCAM mRNAs differ in stability or translation efficiency. In addition post-translational modifications represent a major aspect of these glycoproteins. NCAM is both phosphorylated and sulfated in its transit through the Golgi apparatus (Lyles et al., 1984b). No significant alterations in phosphorylation are thought to accompany nervous tissue development, and it would appear that they are unlikely to have major functional significance. Moreover, activators of protein kinase C, which is involved in the phosphorylation of many proteins, stimulates the synthesis of certain members of the immunoglobulin supergene family but do not modify NCAM synthesis (Doherty et al., 1988). In contrast the level of sulfation decreases during development (Lyles et al., 1984b; Linnemann et al., 1985).

The post-translational modification which has received most attention concerns glycosylation. Expression of the HNK-1 carbohydrate epitope and more recently another carbohydrate epitope called L3 have been studied (Kruse et al., 1984, 1985; Keilhauer et al., 1985; Kücherer et al., 1987) but the degree of sialylation of NCAM has been investigated in most detail. From a functional viewpoint, since the sialic acid content profoundly affects the adhesive binding capacity of NCAM, such modifications assume as much importance as NCAM membrane density. Sialylation of NCAM is unusual in as far as forms of NCAM with high sialic acid levels (as much as 30% of the molecule) contain this sugar in the form of the α-2,8 polysialic acid (PSA, Cunningham et al., 1983; Finne et al., 1983). At the very earliest stages of embryonic development PSA levels appear to be low. Then NCAM with high PSA content is produced, and subsequently PSA content is reduced as development proceeds but certain zones with high PSA NCAM persist in the adult (Sunshine et al., 1987; Aaron and Chesselet, 1989; Rougon et al., 1990). Since the appearance of the three principal NCAM isoforms is not simultaneous during development (NCAM 120 appears somewhat later than the other two) it is not clear if all forms are poly-sialylated to the same degree. Data from explant cultures of mouse and chick tissues suggest that the conversion from high to low PSA NCAM is not due to processing of the embryonic into the adult form, but by *de novo* synthesis of NCAM with lower sialic acid content and that this is due to intracellular regulation of sialyl transferase (Friedlander et al., 1985). More recently (Breen et al., 1987, 1988) found that an endogenous Golgi sialyltransferase was capable of sialylating NCAM 180 and NCAM 120 *in situ*, though curiously obtained no evidence of NCAM 140 sialylation under their conditions. In addition highest tissue activities of this enzyme coincide with highest tissue levels of NCAM PSA.

In pathological tissues increased PSA NCAM levels (Daniloff et al., 1986a) has been linked to the process of cellular reorganisation. In contrast in developing cerebellum no correlation was found between reduction in PSA levels and the migratory status of cerebellar granule neurons (Nagata and Schachner, 1986).

At present no direct data exist on NCAM sialylation in neuroendocrine tissues. In the adult it is likely to be relatively low since immunoblots reveal the presence of discrete NCAM bands. Some data suggest that in the neurohypophysis PSA levels may be higher (Langley et al., 1989a).

4.6 Concluding Remarks on NCAM as a Neuronal and Neuroendocrine Cell Marker

The original concept that NCAM is confined in the adult to neurons and moreover with restricted localization to specialized regions of the neuronal membrane has been gradually eroded over the past few years. Clearcut evidence has demonstrated its expression by other neural cells in both PNS and CNS and also by endocrine cells independant of whether their embryological origin is neuroectodermal or not. In addition the presence of NCAM in adult muscle and certain other non-nervous tissues of adult chicken including skin and lung confirms the wider than neural distribution of this molecule. The cellular expression of NCAM cannot thus be taken as an absolute criterion of neuronal character. NCAM 140 may be regarded as the more typical isoform of neuroendocrine cells and may provide an additional parameter to characterize putative neuroendocrine cells in non-nervous tissues. For example the expression of NCAM 140 by SCLC contrasts with its absence in other lung tumors such as mesotheliomas or adenomas, confirming the endocrine nature of this tumor.

In contrast NCAM 180 appears to be more characteristic of post-mitotic mature neurones than of other neural cells in their terminally differentiated state. Strictly non-neural cells do not appear to express this isoform, although the capacity of paraneurons to synthesize NCAM 180 does not seem to be permanently lost. As has been shown for pheochromocytoma PC12 cells, stimuli which predispose differentiation towards a neuronal phenotype may up-regulate synthesis of NCAM 180. In quantitative terms, however, the expression of NCAM is heavily biased towards the nervous system. Levels found even in neural crest-derived tissues are much lower than those of the CNS. In addition, levels of NCAM have been found to vary between different types of endocrine tumor. For instance, human prolactinomas appear to contain less than growth hormone producing or inactive adenomas of the hypophysis. This may itself be of use to distinguish between tumors particularly if the prolactinoma is inactive.

Nevertheless it is clear that the use of NCAM as a marker for neural and neuroendocrine cells should be treated with caution. No single marker protein fulfills the criterion of absolute specificity. Together with a battery of other marker proteins NCAM isoforms can aid cellular characterization and diagnosis of tumors. Polysialylation of NCAM polypeptides represents an unusual post-translational modification probably unique to members of this family of proteins. Its reappearance or continued presence in adult tissues seems to be associated with either altered intercellular contact phenomena following a pathological process or with the need to retain a higher degree of plasticity in cell contacts in normal tissues. Its presence in Wilms tumor suggests it would be a good marker in this tissue and prompts its search in other tumors. A recent report (Moolenaar et al., 1990) showed the presence of high levels of NCAM sialylation in a cell line derived from an endocrine lung tumor. It is possible that high PSA NCAM forms may be generally associated with the dedifferentiated state of malignantly transformed neural and neuroendocrine cells.

Acknowledgements

Studies from authors' laboratories were supported by Deutsche Krebshilfe, the Deutsche Forschungsgemeinschaft (Gr 681), the C.N.R.S. and I.N.S.E.R.M. and Landesforschungsschwerpunkt No. 32. The authors thank Mrs. B. Mader for the preparation of this manuscript.

4.7 References

1 Aaron, L.I., and Chesselet, M.-F. (1989) Heterogeneous distribution of polysialy-lated neuronal-cell adhesion molecule during post-natal development and in the adult: An immunohistochemical study in the rat brain. *Neurosci.* **28**, 701–710

2 Abo,T., and Balch, C.M. (1981) A differentiation antigen of human NK and K cells identified by a monoclonal antibody (HNK-1). *J. Immunol.* **127**, 1024–1029

3 Aletsee-Ufrecht, M.C., Langley, O.K., Ahnert-Hilger, G., and Gratzl, M. (1988) Cell contact and membrane fusion in endocrine cells. *Studia Biophys.* **127**, 83–88

4 Aletsee-Ufrecht, M.C., Langley, O.K., Rotsch, M., Havemann, K., and Gratzl, M. (1990a) NCAM: A surface marker for human small cell lung cancer cells. *FEBS Letters* **267**, 295–300

5 Aletsee-Ufrecht, M.C., Langley, O.K., Gratzl, O., and Gratzl, M. (1990b) Differential expression of the neural cell adhesion molecule NCAM 140 in human pituitary tumors. *FEBS Let.*, in press

6 Barbas, J.A., Chaix, J.-C., Steinmetz, M., and Goridis, C. (1988) Differential splicing and alternative polyadenylation generates distinct NCAM transcripts and proteins in the mouse. *EMBO J.* **7**, 625–632

7 Barthels, D., Santoni, M.-J., Wille, W., Ruppert, C., Chaix, J.-C., Hirsch, M.-R., Fontecilla-Camps, J.C., and Goridis, C. (1987) Isolation and nucleotide sequence of mouse NCAM cDNA that codes for a Mr 79000 polypeptide without a membrane-spanning region. *EMBO J.* **6**, 907–914

8 Becker, J.W., Erickson, H.P., Hoffman, S., Cunningham, B.A., and Edelman, G.M. (1989) Topology of cell adhesion molecules. *Proc. Natl. Acad. Sci. USA* **86**, 1088–1092

9 Bhat, S., and Silberberg, D.H. (1985) Rat oligodendrocytes have cell adhesion molecules. *Dev. Brain Res.* **19**, 139–145

10 Bhat, S., and Silberberg, D.H. (1988) Developmental expression of neural cell adhesion molecules of oligodendrocytes in vivo and in culture. *J. Neurochem.* **50**, 1830–1838

11 Bock, E. (1972) Identification and characterization of water soluble rat brain antigens. 1. Brain and species specificity. *J. Neurochem.* **19**, 1731–1736

12 Bock, E., Berezin, V., and Rasmussen, S. (1983) Characterization of the D2-cell adhesion molecule, in: *Protides of the Biological Fluids (H. Peeters, ed.) Pergamon,* pp. 75–78

13 Bock, E., Edvardsen, K., Gibson, A., Linnemann, D., Lyles, J.M., and Nybroe, O. (1987) Characterization of soluble forms of NCAM. *FEBS Lett.* **225**, 33–36

14 Bock, E., Jørgensen, O.S., Dittmann, L., and Eng, L.F. (1975) Determination of brain-specific antigens in short term cultivated rat astroglial cells and in rat synaptosomes. *J. Neurochem.* **25**, 867–870

15 Bock, E., Jørgensen, O.S., and Morris, S.T. (1974) Antigen-antibody crossed electrophoresis of rat synaptosomes and synaptic vesicles: Correlation to water-soluble antigens from rat brain. *J. Neurochem.* **22**, 1013–1017

16 Bock, E., Richter-Landsberg, C., Faissner, A., and Schachner, M. (1985) Demonstration of immunochemical identity between the nerve growth factor inducible large external (NILE) glycoprotein and the cell adhesion molecule L1. *EMBO J.* **4**, 2765–2768

17 Bock, E., Yavin, Z., Jørgensen, O.S., and Yavin, E. (1980) Nervous system-specific proteins in developing rat cerebral cells in culture. *J. Neurochem.* **35**, 1297–1302

18 Brackenbury, R., Thiery, J.-P., Rutishauser, U., and Edelman, G.M. (1977) Adhesion among neural cells of the chick embryo. I. An immunological assay for molecules involved in cell-cell binding. *J. Biol. Chem.* **252**, 6835–6840

19 Breen, K.C., Kelly, P.G., and Regan, C.M. (1987) Postnatal D2-CAM/N-CAM sialylation state is controlled by a developmentally regulated golgi sialyltransferase. *J. Neurochem.* **48**, 1486–1493

20 Breen, K.C., and Regan, C.M. (1988) Developmental control of N-CAM sialylation state by Golgi sialyltransferase isoforms. *Development* **104**, 147–154

21 Chuong, C.-M., Crossin, K.L., and Edelman, G.M. (1987) Sequential expression and differential function of multiple adhesion molecules during the formation of cerebellar cortical layers. *J. Cell Biol.* **104**, 331–342

22 Chuong, C.-M., and Edelman, G.M. (1984) Alterations in neural cell adhesion molecules during development of different regions of the nervous system. *J. Neurosci.* **4**, 2354–2368

23 Chuong, C.-M., and Edelman, G.M. (1985a) Expression of cell-adhesion molecules in embryonic induction. 1. Morphogenesis of nestling feathers. *J. Cell Biol.* **101**, 1009–1026

24 Chuong, C.-M., and Edelman, G.M. (1985b) Expression of cell-adhesion molecules in embryonic induction. 2. Morphogenesis of adult feathers. *J. Cell Biol.* **101**, 1027–1043

25 Cole, G.J., and Akeson, R. (1989) Identification of a heparin binding domain of the neural cell adhesion molecule N-CAM using synthetic peptides. *Neuron* **2**, 1157–1165

26 Cole, G.J., and Glaser, L. (1986) A heparin-binding domain from N-CAM is involved in neural cell-substratum adhesion. *J. Cell Biol.* **102**, 403–412

27 Cole, G.J., Loewy, A., Cross, N.V., Akeson, R., and Glaser, L. (1986a) Topographic localization of the heparin-binding domain of the neural cell adhesion molecule N-CAM. *J. Cell Biol.* **103**, 1739–1744

28 Cole, G.J., Loewy, A., and Glaser, L. (1986b) Neuronal cell-cell adhesion depends on interactions of N-CAM with heparin-like molecules. *Nature* **320**, 445–447

29 Coons, A.H. (1961) The beginnings of immunofluorescence. *J. Immunol.* **87**, 499–503

30 Coons, A.H., Creech, H.J., and Jones, R.N. (1941) Immunological properties of an antibody containing a fluorescent group. *Proc. Soc. Exp. Biol.* **47**, 200–202

31 Couly, G.F., and Le Douarin, N.M. (1985) Mapping of the early neural primordium in quail-chick chimeras. I. Developmental relationships between placodes, facial ectoderm and prosencephalon. *Dev. Biol.* **110**, 422–439

32 Couly, G.F., and Le Douarin, N.M. (1987) Mapping of the early neural primordium in quail-chick chimeras. II. The prosencephalic neural plate and neural folds: Implications for the genesis of cephalic human congenital abnormalities. *Dev. Biol.* **120**, 198–214

33 Covault, J., Merlie, J.P., Goridis, C., and Sanes, J.R. (1986) Molecular forms of N-CAM and its RNA in developing and denervated skeletal muscle. *J. Cell Biol.* **102**, 731–739

34 Covault, J. and Sanes, J.R. (1986) Distribution of N-CAM in synaptic and extrasynaptic portions of developing and adult skeletal muscle. *J. Cell Biol.* **102**, 716–730

35 Crossin, K.L., Chuong, C.-M., and Edelman, G.M. (1985) Expression sequences of cell adhesion molecules. *Proc. Natl. Acad. Sci. USA* **82**, 6942–6946
36 Cunningham, B.A. (1986) Cell adhesion molecules: A new perspective on molecular embryology. *Trends Biochem. Sci.* **11**, 423–426
37 Cunningham, B.A., Hemperly, J.J., Murray, B.A., Prediger, E.A., Brackenbury, R., and Edelman G.M. (1987) Neural cell adhesion molecule: Structure immuno-globulin-like domains, cell surface modulation, and alternative RNA splicing. *Science* **236**, 799–806
38 Cunningham, B.A., Hoffman, S., Rutishauser, U., Hemperly, J.J., and Edelman, G.M. (1983) Molecular topography of the neural cell adhesion molecule NCAM: Surface orientation and location of sialic acid-rich and binding regions. *Proc. Natl. Acad. Sci. USA* **80**, 3116–3120
39 Dahl, D., and Bignami, A. (1973) Glial fibrillary acidic protein from normal human brain. Purification and properties. *Brain Res.* **57**, 343–360
40 Daniloff, J.K., Levi, G., Grumet, M., Rieger, F., and Edelman, G.M. (1986a) Altered expression of neuronal cell adhesion molecules induced by nerve injury and repair. *J. Cell Biol.* **103**, 929–945
41 Daniloff, J.K., Chuong, C.-M., Levi, G., and Edelman, G.M. (1986b) Differential distribution of cell adhesion molecules during histogenesis of the chick nervous system. *J. Neurosci.* **6**, 739–758
42 Doherty, P., Mann, D.A., and Walsh, F.S. (1988) Comparison of the effects of NGF, activators of protein kinase C, and a calcium ionophore on the expression of Thy-1 and N-CAM in PC12 cell cultures. *J. Cell Biol.* **107**, 333–340
43 Doherty, P., and Walsh, F.S. (1987) Control of Thy-1 glycoprotein expression in cultures of PC12 cells. *J. Neurochem.* **49**, 610–616
44 Duband, J.-L., Rocher, S., Chen, W.-T., Yamada, K.-M., and Thiery, J.P. (1986) Cell adhesion and migration in the early vertebrate embryo: Location and possible role of the putative fibronectin-receptor complex. *J. Cell Biol.* **102**, 160–178
45 Edelman, G.M. (1983) Cell adhesion molecules. *Science* **219**, 450–457
46 Edelman, G.M. (1984a) Expression of cell adhesion molecules during embryogenesis and regeneration. *Exp. Cell Res.* **161**, 1–16
47 Edelman, G.M. (1984b) Cell-surface modulation and marker multiplicity in neural patterning. *Trends Neurosci.*, **7/3**, 78–84
48 Eng, L.F., and Bigbee, J.W. (1978) Immunohistochemistry of nervous system-specific antigens, in: *Advances in Neurochemistry* (Agranoff, B.W., Aprison, M.H., eds.) Plenum Press, New York, Vol. 3, pp. 43–97
49 Faissner, A., Kruse, J., Goridis, C., Bock, E., and Schachner, M. (1984) The neural cell adhesion molecule L1 is distinct from the N-CAM related group of surface antigens BSP-2 and D2. *EMBO J.* **3**, 733–737
50 Ffrench-Constant, C., Miller, R.H., Kruse, J., Schachner, M., and Raff, M. (1986) Molecular specialization of astrocyte processes at nodes of Ranvier in optic nerve. *J. Cell Biol.* **102**, 844–852
51 Finne, J., Finne, W., Deagostini-Bazin, H., and Goridis, C. (1983) Occurrence of alpha-2–8 linked polysialosyl units in a neural cell adhesion molecules. *Biochem. Biophys. Res. Commun.* **112**, 482–487
52 Frelinger III, A.L., and Rutishauser, U. (1986) Topography of N-CAM structural and functional determinants. II. Placement of monoclonal antibody epitopes. *J. Cell Biol.* **103**, 1729–1737

53 Friedlander, D.R., Brackenbury, R., and Edelman, G.M. (1985) Conversion of embryonic form to adult forms of NCAM in vitro: Results from de novo synthesis of adult forms. *J. Cell Biol.* **101**, 412–419

54 Friedlander, D.R., Grumet, M., and Edelman, G.M. (1986) Nerve growth factor enhances expression of neuron-glia cell adhesion molecule in PC12 cells. *J. Cell Biol.* **102**, 413–419

55 Friedlander, D.R., Mege, R.-M., Cunningham, B.A., and Edelman, G.M. (1989) Cell sorting-out is modulated by both the specificity and amount of different cell adhesion molecules (CAMs) expressed on cell surfaces. *Proc. Natl. Acad. Sci. USA* **86**, 7043–7047

56 Fujita, T. (1977) Concept of paraneurons. *Arch. Histol. Jpn.* **40**, (Suppl.) 1–12

57 Garner, J.A., Watanabe, M., and Rutishauser, U. (1986) Rapid axonal transport of the neural cell adhesion molecule. *J. Neurosci.* **6**, 3242–3249

58 Gatchalian, C.L., Schachner, M., and Sanes, J.R. (1989) Fibroblasts that proliferate near denervated synaptic site in skeletal muscle synthesize the adhesive molecules tenascin (JI), NCAM, fibronectin, and a heparan sulfate proteoglycan. *J. Cell Biol.* **108**, 1873–1890

59 Gennarini, G., Hirn, M., Deagostini-Bazin, H., and Goridis, C. (1984a) Studies on the transmembrane disposition of the neural cell adhesion molecule N-CAM. The use of liposome-inserted radioiodinated N-CAM to study its transbilayer orientation. *Eur. J. Biochem.* **142**, 65–73

60 Gennarini, G., Hirsch, M.-R., He, H.-T., Hirn, M., Finne, J., and Goridis, C. (1986) Differential expression of mouse neural cell-adhesion molecule (N-CAM) mRNA species during brain development and in neural cell lines. *J. Neurosci.* **6**, 1983–1990

61 Gennarini, G., Rougon, G., Deagostini-Bazin, H., Hirn, M., and Goridis, C. (1984b) Studies on the transmembrane disposition of the neural cell adhesion molecule N-CAM. A monoclonal antibody recognizing a cytoplasmic domain and evidence for the presence of phosphoserine residues. *Eur. J. Biochem.* **142**, 57–64

62 Goridis, C., and Wille, W. (1988) The three size classes of mouse NCAM proteins arise from a single gene by a combination of alternative splicing and use of different polyadenylation sites. *Neurochem. Int.* **12**, 269–272

63 Grant, N.J., Demeneix, B., Aunis, D., and Langley, O.K. (1988) Induction of neurofilament phosphorylation in cultured chromaffin cells. *Neuroscience* **27**, 717–726

64 Grumet, M., Rutishauser, U., and Edelman, G.M. (1982) Neural cell adhesion molecule is on embryonic muscle cells and mediates adhesion to nerve cells in vitro. *Nature* **295**, 693–695

65 He, H.-T., Finne, J., and Goridis, C. (1987) Biosynthesis, membrane association, and release of N-CAM 120, a phosphatidylinositol-linked form of the neural cell adhesion molecule. *J. Cell Biol.* **105**, 2489–2500

66 Hirn, M., Ghandour, M.S., Deagostini-Bazin, H., and Goridis, C. (1983) Molecular heterogeneity and structural evolution during cerebellar ontogeny detected by monoclonal antibody of the mouse cell surface antigen BSP-2. *Brain Res.* **265**, 87–100

67 Hirn, M., Pierres, M., Deagostini-Bazin, H., Hirsch, M., and Goridis, C. (1981) Monoclonal antibody against cell surface glycoprotein of neurons. *Brain Res.* **214**, 433–439

68 Hoffman, S., Friedlander, D.R., Chuong, C.-M., Grumet, M., and Edelman, G.M. (1986) Differential contributions of Ng-CAM and N-CAM to cell adhesion in different neural regions. *J. Cell Biol.* **103**, 145–158

69 Hökfelt, T., Johansson, O., Ljungdahl, A., Lundbery, J.M., and Schultzberg, M. (1980) Peptidergic neurons. *Nature* **284**, 515–521

70 Jacobson, M., and Rutishauser, U. (1986) Induction of neural cell adhesion molecule (NCAM) in Xenopus embryos. *Dev. Biol.* **116**, 524–531

71 Jørgensen, O.S., and Bock, E. (1974) Brain specific synaptosomal membrane proteins demonstrated by crossed immunoelectrophoresis. *J. Neurochem.* **23**, 879–880

72 Jørgensen, O.S., Dalouvee, A., Thiery, J.-P., and Edelman, G.M. (1980) The nervous system specific protein D2 is involved in adhesion among neurites from cultured rat ganglia. *FEBS Lett.* **111**, 39–42

73 Jørgensen, O.S., and Honegger, P. (1983) A neuronal adhesion molecule in the central nervous system: The D2-protein. in: *Protides of the Biological Fluids* (H. Peeters, ed.) Pergamon, pp. 79–82

74 Jørgensen, O.S., and Møller, M. (1980) Immunocytochemical demonstration of the D2 protein in the presynaptic complex. *Brain Res.* **194**, 419–429

75 Jørgensen, O.S., and Møller, M. (1981) Immunocytochemical localization of the D2-protein in rat retina. *Brain Res.* **221**, 15–26

76 Jørgensen, O.S., and Møller, M. (1983) A testis antigen related to the brain D2 adhesion protein. *Dev. Biol.* **100**, 275–286

77 Jørgensen, O.S., and Richter-Landsberg, C. (1983) D2-protein in PC12 pheochromocytoma cells after nerve growth factor stimulation. *Neuroscience* **9**, 665–672

78 Keilhauer, G., Faissner, A., and Schachner, M. (1985) Differential inhibition of neurone-neurone, neurone-astrocyte and astrocyte-astrocyte adhesion by L1, L2 and NCAM antibodies. *Nature* **316**, 728–730

79 Kruse, J., Keilhauer, G., Faissner, A., Timpl, R., and Schachner, M. (1985) The J1 glycoprotein – a novel nervous system cell adhesion molecule of the L2/HNK-1 family. *Nature* **316**, 146–148

80 Kruse, J., Mailhammer, R., Wernecke, H., Faissner, A., Sommer, I., Goridis, C., and Schachner, M. (1984) Neural cell adhesion molecules and myelin-associated glycoprotein share a common carbohydrate moiety recognized by monoclonal antibodies L2 and HNK-1. *Nature* **311**, 153–155

81 Kücherer, A., Faissner, A., and Schachner, M. (1987) The novel carbohydrate epitope L3 is shared by some neural cell adhesion molecules. *J. Cell Biol.* **104**, 1597–1602

82 Künemund, V., Jungalwala, F.B., Fischer, G., Chou, D.K.H., Keilhauer, G., and Schachner, M. (1988) The L2/HNK-1 carbohydrate of neural cell adhesion molecules is involved in cell interactions. *J. Cell Biol.* **106**, 213–223

83 Langley, O.K. (1979) Histochemistry of polyanions in peripheral nerve. in: *Complex Carbohydrates of Nervous Tissue* (Margolis, R.U., and Margolis, R.K., eds.) Plenum Press, New York, pp. 193–208

84 Langley, O.K., Aletsee, M.C., and Gratzl, M. (1987) Endocrine cells share expression of N-CAM with neurones. *FEBS Lett.* **220**, 108–112

85 Langley, O.K., Aletsee-Ufrecht, M.C., Grant, N., and Gratzl, M. (1989a) Expression of the neural cell adhesion molecule NCAM in endocrine cells. *J. Histochem. Cytochem.* **37**, 781–791

86 Langley, O.K., Aletsee-Ufrecht, M.C., and Gratzl, M. (1989b) Comparison of the expression of two cell adhesion molecules in endocrine tissues. *J. Neurochem.* **52**, (Suppl.) S 133

87 Langley, O.K., and Ambrose, E.J. (1964) Isolation of a mucopeptide from the surface of Ehrlich ascites tumour cells. *Nature* **204**, 53–54

88 Langley, O.K., and Aunis, D. (1984) Ultrastructural immunocytochemical demonstration of D2-protein in adrenal medulla. *Cell Tissue Res.* **238**, 497–502

89 Langley, O.K., and Aunis, D. (1986) Surface expression of neural cell adhesion molecule in cultured bovine paraneurones: Immunogold and immunoperoxidase methods compared. *Neurosci. Lett.* **64**, 151–156

90 Langley, O.K., Ghandour, M.S., and Gombos, G. (1984) Immunohistochemistry of cell markers in the central nervous system, in: *Handbook of Neurochemistry* (Lajtha, A., ed.) Plenum Press, New York, Vol. 7, pp. 545–611

91 Langley, O.K., Ghandour, M.S., Gombos, G., Hirn, M., and Goridis, C. (1982a) Monoclonal antibodies as neural cell surface markers. *Neurochem. Res.* **7**, 349–362

92 Langley, O.K., Gombos, G., Hirn, M., and Goridis, C. (1983) Distribution of the neural antigen BSP-2 in the cerebellum during development. *Int. J. Devl. Neurosci.* **6**, 393–401

93 Langley, O.K., Reeber, A., Vincendon, G., and Zanetta, J.-P. (1982b) Fine structural localization of a new Purkinje cell-specific glycoprotein subunit. Immunoelectron microscopical study. *J. Comp. Neurol.* **208**, 335–344

94 Le Douarin, N.M. (1978) The embryological origins of the endocrine cells associated with the digestive tract: Experimental analysis based on the use of a stable marking technique, in: *Gut Hormones (Bloom, S.F., ed.) Churchill, London, pp. 49–56*

95 Levi, G., Crossin, K.L., and Edelman, G.M. (1987) Expression sequences and distribution of two primary cell adhesion molecules during embryonic development of Xenopus laevis. *J. Cell Biol.* **105**, 2359–2372

96 Linnemann, D., Lyles, J.M., and Bock, E. (1985) A developmental study of the biosynthesis of the neural cell adhesion molecule. *Dev. Neurosci.* **7**, 230–238

97 Lohmann, S.M., Walter, U., Miller, P.E., Greengard, P., and de Camilli, P. (1981) Immunohistochemical localization of cyclic GMP-dependent protein kinase in mammalian brain. *Proc. Natl. Acad. Sci. USA* **78**, 653–657

98 Lyles, J.M., Linnemann, D., and Bock, E. (1984a) Biosynthesis of the D2-cell adhesion molecule: Post-translational modifications, intracellular transport, and developmental changes. *J. Cell Biol.* **99**, 2082–2091

99 Lyles, J.M., Norrid, B., and Bock, E. (1984b) Biosynthesis of the D2-cell adhesion molecule. Pulse chase studies in fetal rat neuronal cells. *J. Cell Biol.* **98**, 2077–2081

100 McGuire, J., Greene, L., and Furano, A. (1978) NGF stimulates incorporation of fucose or glucosamine into an external glycoprotein in cultured rat PC12 pheochromocytoma cells. *Cell* **15**, 357–365

101 Maier, C.E., Watanabe, M., Singor, M., McQuarrie, I.G., Sunshine, J., and Rutishauser, U. (1986) Expression and function of neural cell adhesion molecule during limb regeneration. *Proc. Natl. Acad. Sci. USA* **83**, 8395–8399

102 Martini, R., and Schachner, M. (1988) Immunoelectron microscopic localization of neural cell adhesion molecules (L1, N-CAM, and myelin-associated glycoprotein) in regenerating adult mouse sciatic nerve. *J. Cell Biol.* **106**, 1735–1746

103 Matus, A., and Mughal, S. (1975) Immunohistochemical localization of S100 protein in brain. *Nature* **258**, 746–748

104 Mikoshiba, K., Huchet, M., and Changeux, H.-P. (1979) Biochemical and immunological studies on the P400 protein, a protein characteristic of the Purkinje cell from mouse and rat cerebellum. *Dev. Neurosci.* **2**, 254–275

105 Mirsky, R., Jessen, K.R., Schachner, M., and Goridis, C. (1986) Distribution of the adhesion molecules N-CAM and L1 on peripheral neurons and glia in adult rats. *J. Neurocytol.* **15**, 799–815

106 Moolenaar, C.E.C.K., Muller, E.J., Schol, D.J., Figdor, C.G., Bock, E., Bitter-Suermann, D., and Michalides, R.J.A.M. (1990) Expression of neural cell adhesion molecule-related sialoglycoprotein in small cell lung cancer and neuroblastoma cell lines H69 and CHP-212. *Cancer Res.* **50**, 1102–1106

107 Moore, S.E., Thompson, J., Kirkness, V., Dickon, J.G., and Walsh, F.S. (1987) Skeletal muscle neural cell adhesion molecule (N-CAM): Changes in protein and mRNA species during myogenesis of muscle cell lines. *J. Cell Biol.* **105**, 1377–1386

108 Murray, B.A., Owens, G.C., Prediger, E.A., Crossin, K.L., Cunningham, B.A., and Edelman, G.M. (1986) Cell surface modulation of the neural cell adhesion molecule resulting from alternative mRNA splicing in a tissue-specific developmental sequence. *J. Cell Biol.* **103**, 1431–1439

109 Nagata, I., and Schachner, M. (1986) Conversion of embryonic to adult form of the neural cell adhesion molecule (N-CAM) does not correlate with pre- and postmigratory states of mouse cerebellar granule neurons. *Neurosci. Lett.* **63**, 153–158

110 Nègre-Aminou, P., Massacrier, A., Hirn, M., and Cau, P. (1988) Quantitative analysis of rat brain neurons developing in primary cultures. II. Changes in the distribution of N-CAM associated to neuronal cell surfaces. *Devl. Brain Res.* **40**, 171–180

111 Noble, M., Albrechtsen, M., Mëller, C., Lyles, J., Bock, E., Goridis, C.,Watanabe, M., and Rutishauser, U. (1985) Glial cells express N-CAM/D2-CAM-like polypeptides in vitro. *Nature* **316**, 725–728

112 Noronha, A., Ilyas, A., Quarles, R.H., Antonicek, H., and Schachner, M. (1986) Molecular specificity of L2 monoclonal antibodies that bind to carbohydrate determinants of neural cell adhesion molecules and resemble other monoclonal antibodies recognising the myelin-associated glycoprotein. *Brain Res.* **385**, 237–244

113 Nybroe O., Albrechtsen, M., Dahlin, J., Linneman, D., Lyles, J.M., Mëller, C.J., and Bock, E. (1985) Biosynthesis of the neural cell adhesion molecule. Characterization of polypeptide C. *J. Cell Biol.* **101**, 2310–2315

114 Nybroe, O., Gibson, A., Mëller, C.J., Rhode, H., Dahlin, J., and Bock, E. (1986) Expression of N-CAM polypeptides in neurons. *Neurochem. Int.* **9**, 539–544

115 Nybroe, O., Linnemann, D., and Bock, E. (1988) NCAM biosynthesis in brain. *Neurochem. Int.* **12**, 251–262

116 Nybroe, O., Moran, N., and Bock, E. (1989) Equilibrium binding analysis of neural cell adhesion molecule binding to heparin. *J. Neurochem.* **52**, 1947–1949

117 Owens, G.C., Edelman, G.M., and Cunningham, B.A. (1987) Organization of the neural cell adhesion molecule (N-CAM) gene: Alternative exon usage as the basis for different membrane-associated domains. *Proc. Natl. Acad. Sci. USA* **84**, 294–298

118 Palay, S.L., and Chan-Palay, V. (1974) *Cerebellar cortex*. Springer, Berlin, pp. 11–62

119 Patel, K., Moore, S.E., Dickson, G., Rossell, R.J., Beverley, P.C., Kemshead, J.T., and Walsh, F.S. (1989) Neural cell adhesion molecule (NCAM) is the antigen recognized by monoclonal antibodies of similar specificity in small-cell lung carcinoma and neuroblastoma. *Int. J. Cancer* **44**, 573–578

120 Pearse, A.G.E. (1968) Common cytochemical and ultrastructural characteristics of cells producing polypeptide hormones (the APUD series) and their relevance to thyroid and ultimobranchial C cells and calcitonin. *Proc. Roy. Soc. B* **170**, 71–80

121 Pearse, A.G.E. (1969) The cytochemistry and ultrastructure of polypeptide hormone-producing cells of the APUD series and the embryologic, physiologic and pathologic implications of the concept. *J. Histochem. Cytochem.* **17**, 303–313

122 Pearse, A.G.E. (1983) Islet development and the APUD concept, in: *Pancreatic Pathology* (Klöppel, G., and Heitz, P.U., eds.) Churchill Livingstone, Edinburgh, pp. 125–132

123 Pigott, R., and Kelly, J.S. (1986) Immunocytochemical and biochemical studies with the monoclonal antibody 69A1: Similarities of the antigen with cell adhesion molecules L1, NILE and Ng-CAM. *Devl. Brain Res.* **29**, 111–122

124 Pollerberg, G.E., Burridge, K., Krebs, K.E., Goodman, S.R., and Schachner, M. (1987) The 180-kD component of the neural cell adhesion molecule N-CAM is involved in cell-cell contacts and cytoskeleton-membrane interactions. *Cell Tissue Res.* **250**, 227–236

125 Pollerberg, E.G., Sadoul, R., Goridis, C., and Schachner, M. (1985) Selective expression of the 180-kD component of the neural cell adhesion molecule N-CAM during development. *J. Cell Biol.* **101**, 1921–1929

126 Pollerberg, G.E., Schachner, M., and Davoust, J. (1986) Differentiation state-dependent surface mobilities of two forms of the neural cell adhesion molecule. *Nature* **324**, 462–465

127 Poltorak, M., Sadoul, R., Keilhauer, G., Lauda, C., and Schachner, M. (1987) The myelin-associated glycoprotein (MAG), a member of the L2/HNK-1 family of neural cell adhesion molecules, is involved in neuron-oligodendrocyte and oligodendrocyte-oligodendrocyte interaction. *J. Cell Biol.* **105**, 1893–1899

128 Prentice, H.M., Moore, S.E., Dickson, J.G., Doherty, P., and Walsh, F.S. (1987) Nerve growth factor-induced changes in neural cell adhesion molecule (N-CAM) in PC12 cells. *EMBO J.* **6**, 1859–1863

129 Prieto, A.L., Crossin, K.L., Cunningham, B.A., and Edelman, G.M. (1989) Localization of mRNA for neural cell adhesion molecule (N-CAM) polypeptides in neural and non neural tissues by in situ hybridization. *Proc. Natl. Acad. Sci. USA* **86**, 9579–9583

130 Ramon Y Cajal, S. (1905) Genesis de la fibras nerviosa del embryion y observaciones contrarias a la teoria catenaria. *Trab. Lab. Invest. Biol.* **4**, 277–294

131 Reeber, A., Vincendon, G., and Zanetta, J.-P. (1981) Isolation and immunohisto-chemical localization of a Purkinje cell specific glycoprotein subunit from rat cerebellum. *Brain Res.* **229**, 53–65

132 Rieger, F., Daniloff, J.K., Pincon-Raymond, M., Crossin, K.L., Grumet, M., and Edelman, G.M. (1986) Neuronal cell adhesion molecules and cytotactin are colocalized at the node of Ranvier. *J. Cell Biol.* **103**, 379–391

133 Rieger, F., Nicolet, M., Pincon-Raymond, M., Murawsky, M., Levi, G., and Edelman, G.M. (1988) Distribution and role in regeneration of N-CAM in the basal laminae of muscle and Schwann cells. *J. Cell Biol.* **107**, 707–719

134 Roth, J., Baetens, D., Norman, A.W., and Garcia-Segura, L-M. (1981) Specific neurons in chick central nervous system stain with an antibody against chick intestinal vitamin D-dependent calcium-binding protein. *Brain Res.* **222**, 452–457

135 Roth, J., Taatjes, D.J., Bitter-Suermann, D., and Finne, J. (1987) Polysialic acid units are spatially and temporally expressed in developing postnatal rat kidney. *Proc. Natl. Acad. Sci. USA* **84**, 1969–1973

136 Roth, J., Zuber, C., Wagner, P., Taatjes, D.J., Weisgerber, C., Heitz, P.U., Goridis, C., and Bitter-Suermann, D. (1988) Reexpression of poly(sialic acid) units of the neural cell adhesion molecule in Wilms tumor. *Proc. Natl. Acad. Sci. USA* **85**, 2999–3003

137 Rougon, G., Deagostini-Bazin, H., Hirn, M., and Goridis, C. (1982) Tissue- and developmental stage-specific forms of a neural cell surface antigen linked to differences in glycosylation of a common polypeptide. *EMBO J.* **1**, 1239–1244

138 Rougon, G., Dubois, C., Buckley, N., Magnani, J.L., and Zollinger W. (1986) Monoclonal antibody against meningococcus group B polysaccharides distinguishes embryonic from adult N-CAM. *J. Cell Biol.* **103**, 2429–2437

139 Rougon, G., Nedelec, J., Malatert, P., Goridis, C., and Chesselet, M.-F. Post-translational modifications of neural cell surface molecules. *Acta histochem. (in press)*

140 Rutishauser, U. (1984) Developmental biology of a neural cell adhesion molecule. *Nature* **310**, 549–554

141 Rutishauser, U., Acheson, A., Hall, A.K., Mann, D.M., Sunshine, J. (1988) The neural cell adhesion molecule (NCAM) as a regulator of cell-cell interactions. *Science* **240**, 53–57

142 Rutishauser, U., Gall, W.E., and Edelman, G.M. (1978a) Adhesion among neural cells of the chick embryo. IV. Role of the cell surface molecule CAM in the formation of neurite bundles in cultures of spinal ganglia. *J. Cell Biol.* **79**, 382–393

143 Rutishauser, U., and Goridis, C. (1986) NCAM: The molecule and its genetics. *Trends Genet.* **2**, 72–76

144 Rutishauser, U., Thiery, J.-P., Brackenbury, R., and Edelman, G.M. (1978b) Adhesion among neural cells of the chick embryo. III. Relationship of the surface molecule CAM to cell adhesion and the development of histotypic patterns. *J. Cell Biol.* **79**, 371–381

145 Rutishauser, U., Watanabe, M., Silver, J., Troy, F.A., and Vimr, E.R. (1985) Specific alteration of NCAM-mediated cell adhesion by an endoneuraminidase. *J. Cell Biol.* **101**, 1842–1849

146 Sanes, J.R., Schachner, M., and Covault, J. (1986) Expression of several adhesive macromolecules (NCAM, L1, J1, NILE, ovomorulin, laminin, fibronectin, and a heparin sulfate proteoglycan) in embryonic, adult and denervated skeletal muscle. *J. Cell Biol.* **102**, 420–431

147 Santoni, M.-J., Barthels, D., Barbas, J.A., Hirsch, M.-R., Steinmetz, M., Goridis, C., and Wille, W. (1987) Analysis of cDNA clones that code for the transmembrane forms of the mouse neural cell adhesion molecule (NCAM) and are generated by alternative RNA splicing. *Nucleic Acids Res.* **15**, 8627–8647

148 Santoni, M.-J., Barthels, D., Vopper, G., Boned, A., Goridis, C., and Wille, W. (1989) Differentiatial exon usage involving an unusual splicing mechanism generates at least eight types of NCAM cDNA in mouse brain. *EMBO J.* **8**, 385–392

149 Schlosshauer, B. (1989) Purification of neuronal cell surface proteins and generation of epitope-specific monoclonal antibodies against cell adhesion molecules. *J. Neurochem.* **52**, 82–92

150 Schuller-Petrovic, S., Gebhart, W., Lassmann, H., Rumpold, H., and Kraft, D. (1983) A shared antigenic determinant between natural killer cells and nervous tissue. *Nature* **306**, 179–181

151 Seilheimer, B., and Schachner, M. (1988) Studies of adhesion molecules mediating interactions between cells of peripheral nervous system indicate a major role for L1 in mediating sensory neuron growth on Schwann cells in culture. *J. Cell Biol.* **107**, 341–351

152 Seilheimer, B., Persohn, E., and Schachner, M. (1989) Neural cell adhesion molecule expression is regulated by Schwann cell-neuron interactions in culture. *J. Cell Biol.* **108**, 1909–1915

153 Small, S.J., Shull, G.E., Santoni, M.-J., and Akeson, R. (1987) Identification of a cDNA clone that contains the complete coding sequence for a 140-kD rat NCAM polypeptide. *J. Cell Biol.* **105**, 2335–2345

154 Swaab, D.F., Pool, C.W., and van Leeuwen, F.W. (1977) Can specificity ever be proved in immunocytochemical staining? *J. Histochem. Cytochem.* **25**, 388–391

155 Sunshine, J., Balak, K., Rutishauser, U., and Jacobson, M. (1987) Changes in neural cell adhesion molecule (NCAM) structure during vertebrate neural development. *Proc. Natl. Acad. Sci. USA* **84**, 5986–5990

156 Teitelman, G., Lee, J., and Reis, D.J. (1987) Differentiation of prospective mouse pancreatic islet cells during development in vitro and during regeneration. *Dev. Biol.* **120**, 425–433

157 Teitelman, G., and Lee, J.K. (1987) Cell lineage analysis of pancreatic islet cell development: Glucagon and insulin cells arise from catecholaminergic precursors present in the pancreatic duct. *Dev. Biol.* **121**, 454–466

158 Thiery, J.-P., Duband, J.-L., Rutishauser, U., and Edelman, G.M. (1982) Cell adhesion molecules in early chicken embryogenesis. *Proc. Natl. Acad. Sci. USA* **79**, 6737–6741

159 Thor, G., Pollerberg, E.G., and Schachner, M. (1986) Molecular association of two neural cell adhesion molecules within the surface membrane of cultured mouse neuroblastoma cells. *Neurosci. Lett.* **66**, 121–126

160 Trotter, J., Bitter-Suermann, D., and Schachner, M. (1989) Differentiation-regulated loss of the polysialylated embryonic form and expression of the different polypeptides of the neural cell adhesion molecule by cultured oligodendrocytes and myelin. *J. Neurosci. Res.* **22**, 369–383

161 Van den Pol, A.N., di Porzio, U., and Rutishauser, U. (1986) Growth cone localization of neural cell adhesion molecule on central nervous system neurons in vitro. *J. Cell Biol.* **102**, 2281–2294

162 Walsh, F.S. (1988) The N-CAM gene is a complex transcriptional unit. *Neurochem. Int.* **12**, 263–267

163 Williams, A.E. (1987) A year in the life of the immunoglobulin superfamily. *Immunol. Today* **8**, 298–303

164 Williams, R.K., Goridis, C., and Akeson, R. (1985) Individual neural cell types express immunologically distinct N-CAM forms. *J. Cell Biol.* **101**, 36–42
165 Willison, H.J., Minna, J.D., Brady, R.O., and Quarles, R.H. (1986) Glycoconjugates in nervous tissue and small cell lung cancer share immunologically cross-reactive carbohydrate determinants. *J. Neuroimmunol.* **10**, 353–365

Part III

Applications

5 Neuron Specific Enolase as a Clinical Tool in Neurologic and Endocrine Disease

Paul J. Marangos

5.1 Introduction

The characterization of cell specific proteins has become a viable strategy for elucidating the mechanisms involved in differentiated tissue and organ functions. The brain is probably the most differentiated of all organs and certainly represents the most significant challenge for the molecular physiologist with regard to the nature of its function. Theoretically, if the function of all the proteins specific to a given cell type were known then we would fully understand how that cell performs its specific tasks. This was in fact the motivation for the studies performed by Blake Moore and his colleagues during the 1960's when he utilized various methods of protein fractionation to elucidate the soluble proteins specific to brain tissue (Moore and McGregor, 1965; Moore, 1973). Those pioneering studies led to the description of two apparently novel brain proteins which Moore termed the 14-3-2 and the S-100 protein. There were in fact a number of acidic, soluble proteins that only appeared in extracts of the brain and were not readily detected in liver and muscle extracts.

In the several years that followed the initial description of the 14-3-2 and S-100 proteins it was shown by somewhat circumstantial evidence that 14-3-2 was probably a neuronal component and that S-100 was glial (Cicero et al., 1972; Moore, 1973; Cicero et al., 1970).

In the mid-70's it was further shown when specific antibodies to the 14-3-2 proteins were utilized in immunocytochemical staining protocols that this protein was indeed highly localized to neurons and furthermore that it was cytoplasmic (Pickel et al., 1975). The protein at that time was redesignated as **n**euron **s**pecific **p**rotein or NSP (Marangos et al., 1978 Marangos and Zomzely-Neurath, 1976) since this was more descriptive in nature than 14-3-2. Shortly, thereafter it was further demonstrated that NSP was in fact a neuron specific form of the ubiquitous glycolytic enzyme enolase, at which time it was again re-named to **n**euron **s**pecific **e**nolase or NSE (Bock and Dissing, 1975; Marangos et al., 1976; Fletcher et al., 1976). NSE has therefore had a somewhat confusing history over the years. To further complicate matters its structural components (dimeric protein) have given rise to yet another name for NSE.

5.2 Structural and Functional Properties of NSE

NSE has now been purified to essentially homogeneity from a number of species including rat, cat and human brain (Marangos et al., 1975; Scarna et al., 1982). A number of purification procedures have now been published with a wide variation regarding the degree to which the final product had been assessed for purity (Marangos et al., 1975; Soler Federsppiel et al., 1987; Keller et al., 1981). The structural features of NSE are summarized in Tab. 5-1. Like enolases from other tissues, NSE is a dimer which is composed of a subunit that is only synthesized in a restricted number of cell types (neurons and neuroendocrine cells). The properties of the enolase subunit (gamma) are sufficiently unique from that of the liver (alpha) or muscle (beta) enzyme to warrant it being considered a separate and distinct gene product (Marangos et al., 1977). The most convincing evidence derives from the fact that antibodies raised against the gamma subunit (NSE) do not react with either the liver or muscle enzyme (Marangos et al., 1977). It has also been shown that antisera to the alpha subunit does not react with NSE.

Tab. 5-1 Structural Properties of Human Brain NSE.

Molecular weight	78,000
Isoelectric point	5.0
Subunit composition	2 gamma
	(39,000)
Reactivity with anti-NSE serum	+++
Reactivity with anti-liver enolase serum	−

Tab. 5-2 Comparison of NSE and Liver Enolase.

Functional Properties	NSE	Liver Enolase
Substrate K_m	1.2×10^{-4}	1.3×10^{-4}
K_a Mg^{++}	2.4×10^{-4}	6.1×10^{-4}
Temperature stability (55 °C)	+	−
Urea (3M) stability	+	−
Chloride stability (0.5M)	+	−

Molecular biological studies involving the cloning of the NSE gene now have supported previous protein structural and immunological evidence that NSE is in fact a unique gene product (Sakimura et al., 1985 a, b). These studies clearly show a difference in the nucleotide sequence of the gamma gene relative to the alpha gene. It is therefore clear that the NSE isoenzyme does in fact represent a considerable biological investment from the evolutionary perspective and that it is not simply a cell specific isoenzyme that differs by a single derivatized amino acid residue.

Since structure and function are generally related and the structure of NSE appears to be so different from that of the other enolases some effort has been expended to determine whether NSE is functionally different from the enolase isoenzyme present in liver. Kinetically the neuronal enolase appears to be quite similar to the liver enzyme (Tab. 5-2) with some differences in the affinity of the co-factor magnesium observed. The standard kinetic comparison between the isoenzymes is rather unremarkable and difficult to ascribe any physiological significance to. However, the picture becomes somewhat more interesting when other parameters such as stability towards temperature and urea were compared (Tab. 5-2). Here large differences were observed with NSE being much more stable in these protocols (Marangos et al, 1978). The most interesting differences emerged when the respective stabilities of NSE and liver enolase were evaluated in the presence of elevated halogen concentration. Chloride and bromide at millimolar levels dissociates and inactivates liver and yeast enolase (Marangos et al, 1978a) whereas NSE is much more stable towards these conditions. This finding is of considerable interest since one of the relatively unique aspects of neurons is that rather high levels of chloride can accumulate during periods of repeated depolarization. It therefore appears that NSE might have evolved to deal with this aspect of the neuronal milieu since non-neuronal isoenzymes would become inactivated in neuronal cytoplasm under conditions of high functional activity. This would result in an inhibition of glycolysis at the time when it is needed most, i.e. during periods of high energy requirements.

5.3 Cellular Localization of NSE

Several groups have reported on the preparation of both polyclonal (Marangos et al, 1975b) and monoclonal antibodies to NSE (Seshi and Bell, 1985). The quality of data from studies utilizing antibodies as immunocytochemical staining probes is critically dependant on the purity of the NSE that is injected into the host animal and therefore the staining results from different laboratories have varied. Criteria such as lack of reactivity with non-neuronal enolase have only been applied to a limited number of the currently available antibodies (Marangos et al, 1977). Nevertheless, the consensus is fairly compelling that NSE is highly localized to neuronal cytoplasm in mature nervous tissue (Schmechel et al, 1979; Schmechel and Marangos, 1983). Virtually no glial staining is observed and consequently NSE is gradually becoming accepted as a generalized functional marker for nerve cells. There does appear to be a range of staining intensities where various neural subtypes are compared and this may reflect the state of functional activity of the respective cells.

What makes NSE even more valuable as a neural cell marker is the fact that it appears to only be present in differentiated neurons (Schmechel et al, 1980; Whitehead et al, 1982). Developmental studies have now clearly shown that undifferentiated nerve cells actually contain the liver type enolase isoenzyme (alpha-alpha) (Schmechel et al, 1980) and that the NSE gamma subunit only appears after functional synapse formations (Whitehead et al, 1982; Maxwell et al, 1982). NSE therefore is not only useful as a general neural marker but also as an index of differentiation and of considerable value to developmental neurobiologists.

Radioimmunoassays for NSE have been developed in a number of laboratories (Parma et al, 1981; Soler Federsppiel et al, 1987; Kato et al, 1981) in a coordinated fashion with the immunocytochemical protocols mentioned above. These procedures are again only valuable when the antisera employed are highly specific and the antigen utilized to both immunize the rabbit or goat and as the radioligand in the assay is highly pure. Again as with the immunocytochemical studies there has been a considerable variation in the published literature in this regard since a range of standards have been employed in assessing the specificity of the various assays. The most appropriate standard for RIA specificity assessment is the lack of reactivity with purified non-neuronal (liver) enolase (Parma et al, 1981).

Through the utilization of both immunocytochemical and radioimmunological techniques it has been shown that NSE or at least the gamma subunit of enolase is present in a wide assortment of neuroendocrine cells such as adrenal

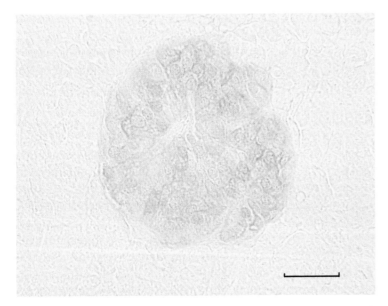

Fig. 5-1 Paraffin section of human pancreas immunostained for NSE. Note the exclusive staining of the islet alls.

medullary chromaffin cells, pineal cells, pancreatic islet cells (Fig. 5-1) and many other cells of the APUD classification (Pearse, 1976; Schmechel et al., 1979). The neuroendocrine cells have somewhat lower levels of NSE than neurons and they also contain non-neural forms of enolase, both of which distinguish them from neurons which have all of their enolase in the form of NSE. Virtually all of the neuroendocrine, APUD cell types that have been evaluated contain NSE immunoreactivity which simply means they have the gamma subunit. Since hybrid forms of enolase do exist (Marangos et al., 1977) it is possible that the gamma subunit present in neuroendocrine cells is in the form of the alpha-gamma hybrid.

5.4 Clinical Applications of the NSE Methodology

Since NSE is localized in neurons and neuroendocrine cells the potential exsits that it could be used as a diagnostic probe for disease states characterized by pathologies involving these cell types. One would expect that in various neurological disorders such as stroke or neurodegenerative disease that NSE levels might accumulate in either cerebrospinal fluid (CSF) or perhaps even in serum. For the pathologist, NSE immunocytochemistry might also be used for the typing of cells. Turnover of the neuro-endocrine cells (APUDomas) might also be expected to result in elevated serum levels of NSE. In fact all of the above approaches have been exploited with the NSE methodology and each is summarized below.

As regards the neurological diseases NSE levels in CSF have been shown to be of value as an index of CNS damage post-stroke (Royds et al., 1981; May et al., 1984). Similar results, although rather preliminary in nature have also been generated in head trauma (Dauberschmidt et al., 1983). NSE determinations in these patient populations may therefore in the future prove to be of some prognostic value. Efforts in the neurodegeneration area have been somewhat more frustrating in that clear correlations between NSE levels and conditions such as Alzheimer's disease and amyotrophic laterosclerosis have thus far proven to be illusive (unpublished data). It may be that a temporal problem exists in these studies such that the pathological insult may occur long (possibly months to years) before the NSE sampling takes place. In this case it may prove quite difficult to utilize serum or CSF NSE levels as a diagnostic agent in the progressive neurodegenerative disease.

Considerably more success has been obtained using serum NSE levels diagnostically in various neuroendocrine cancers. Particularly serum NSE levels are markedly elevated in small cell lung cancer patients (Marangos and Goodwin, 1978; Johnson et al., 1984; Ariyoshi et al., 1983) and in pediatric neuroblastoma patients (Zeltzer et al., 1983; Zeltzer et al., 1985). In both patient populations serum NSE levels can be elevated several hundredfold higher than controls and the levels are correlated with the clinical course of the illness. In neither case does the NSE elevation appear to represent an early detection marker for the illness but it may be an early detection marker for relapse in small cell lung cancer patients (Johnson et al., 1984). It is therefore possible that serial serum NSE determinations in patients experiencing remission may indicate the relapse sooner and therefore make it possible to reinstate salvage chemotherapy sooner. Whether this will alter the course and prognosis of small cell lung cancer remains to be seen.

In pediatric neuroblastoma patients, serum NSE levels are correlated with survival time and in those patients below 1 year of age it was possible to discriminate between survival and deaths (Zeltzer et al., 1985). The serum NSE assay is therefore of considerable use both diagnostically and prognostically in both of these patient populations as well as in some other neuroendrocrine cancers where smaller studies have shown serum NSE elevations (Prinz and Marangos, 1983; Wick et al., 1983). Obviously, it is necessary to have other diagnostic information to differentiate the various neuroendocrine cancers from each other.

Tab. 5-3 Applications of NSE Methodology.

---

1) Functional marker for neurons, neuroendocrine cells as an aid to tissue typing
2) Index of neuronal maturation in developmental studies
3) Diagnostic and prognostic tool (serum, cerebrospinal fluid)
   – stroke and head trauma
   – neuroendocrine cancers
   – small cell lung cancer
   – neuroblastoma

NSE is therefore a rather interesting variant of enolase that has provided us with not only a valuable marker for neurons (in brain) and neuro-endocrine cells (in the periphery) but also with a diagnostic tool that has already been shown to have considerable utility in a number of clinical situations. Tab. 5-3 summarizes some of the applications of the NSE methodology for basic researchers, pathologists, and clinicians. It is therefore likely that NSE histochemistry and serum RIA determination will become useful in the clinical workup of various neuroendocrine neoplasms.

5.5 References

1 Ariyoshi, Y., Kato, K., Ishiguro, Y., Ota, K., Sato, T., Suchi, T. (1983). Evaluation of serum neuron specific enolase as a tumor marker for carcinoma of the lung. *Gann* **74**, 219–225

2 Bock, E., Dissing, J. (1975). Demonstration of enolase activity connected to the brain specific protein 14-3-2. *Scand. J. Immunol.* (Suppl. 2) **4**, 31–36

3 Cicero, T.J., Cowan, W.M., Moore, B.W., Suntzeff, V. (1970). The cellular localization of the two brain specific proteins, S-100 and 14-3-2. *Brain Res.* **18**, 25–34

4 Cicero, T.J., Ferrendelli, J.A., Suntzeff, V., Moore, B.W. (1972). Regional changes in CNS levels of the S-100 and 14-3-2 proteins during development and aging of the mouse. *J. Neurochem.* **19**, 2119–2125

5 Dauberschmidt, R., Marangos, P.J., Zinsmeyer, J., Bender, V., Klages, G., Gross, J. (1983). Severe head trauma and the changes of concentration of neuron specific enolase in plasma and in cerebrospinal fluid. *Clin. Chem. Acta* **131**, 165–170

6 Fletcher, L., Rider, C.C., Taylor, C.B. (1976). Enolase isoenzymes. III. Chromatographic and immunological characteristics of rat brain enolase. *Biochim. Biophys. Acta* **452**, 245–252

7 Hay, E., Royds, J.A., Aelwyn, G., Davies-Jones, B., Lewtas, N.A., Timperly, W.R., Taylor, C.B. (1984). Cerebrospinal fluid enolase in stroke. *J. Neurol. Neurosurg. Psych.* **47**, 724–729

8 Johnson, D.H., Marangos, P.J., Forbes, J.T., Hainsworth, J.D.,Welch, R.V., Hande, K.R., Greco, F.A. (1984). Potential utility of serum neuron specific enolase levels in small cell carcinoma of the lung. *Cancer Res.* **44**, 5409–5411

9 Kato, K., Suzuki, F., Semba, R. (1981). Determination of brain enolase isoenzymes with an enzyme immunoassay at the level of single neurons. *J. Neurochem.* **37**, 998–1005

10 Keller, A., Scarna, H., Mermet, A., Pujol, J.F. (1981). Biochemical and immunological properties of the mouse brain enolase purified by a simple method. *J. Neurochem.* **36**, 1389–1397

11 Marangos, P.J., Goodwin, F.K (1978). Neuron specific protein (NSP) in neuroblastoma cells: Relation to differentiation. *Brain Res.* **145**, 49–58

12 Marangos, P.J., Parma, A.M., Goodwin, F.K. (1978a). Functional properties of neuronal and glial isoenzymes of brain enolase. *J. Neurochem.* **31**, 727–732

13 Marangos, P.J., Zomzely-Neurath, C. (1976). Determination and characterization of neuron specific protein (NSP) associated enolase activity. *Biochem. Biophys. Commun.* **68**, 1309–1316

14 Marangos, P.J., Zomzely-Neurath, C., Goodwin, F.K. (1977). Structural and functional properties of neuron specific protein (NSP) from rat, cat and human brain. *J. Neurochem.* **28**, 1097–1107

15 Marangos, P.J., Zomzely-Neurath, C., Luk, C.M., York, C. (1975). Isolation and characterization of the nervous system specific protein 14-3-2 from rat brain. *J. Biol. Chem.* **250**, 1884–1901

16 Marangos, P.J., Zomzely-Neurath, C.,York, C. (1975b). Immunological studies of a nerve specific protein (NSP). *Arch. Biochem. Biophys.* **170**, 289–293

17 Maxwell, G.D., Whitehead, M.C., Connolly, S.M., Marangos, P.J. (1982). Development of neuron-specific enolase immunoreactivity in avian nervous tissue *in vivo* and *in vitro*. *Dev. Brain. Res.* **3**, 401–419

18 Moore, B.W. (1973). Brain specific proteins. In: *Proteins of the Nervous System*, (D.J. Schneider, ed.) Raven, New York, pp. 1–12

19 Moore, B.W., McGregor, D. (1965). Chromatographic and electrophoretic fractionalation of soluble proteins of brain and liver. *J. Biol. Chem.* **240**, 1647–1653

20 Parma, A.M., Marangos, P.J., Goodwin, F.K. (1981). A more sensitive radioimmunoassay for neuron specific enolase (NSE) suitable for cerebrospinal fluid determinations. *J. Neurochem.* **36**, 1093–1096

21 Pearse, A.G.E. (1976). Neurotransmission and APUD concept. In: *Chromaffin, Enterochromaffin and Related Cells* (R. E. Coupland, T. Fujita eds.). Elsevier, Amsterdam p 142

22 Pickel,V.M., Reis, D.J., Marangos, P.J., Zomzely-Neurath, C. (1975). Immunocytochemical localization of nervous system specific protein (NSP-R) in rat brain. *Brain Res.* **105**, 184–187

23 Prinz, R.A., Marangos, P.J. (1983). Serum neuron specific enolase: A serum marker for non-functioning pancreatic islet cell carcinoma. *Am J. Surg.* **145**, 77–81

24 Royds, J.A., Timperley, W.R., Taylor, C.B. (1981). Levels of enolase and other enzymes in the cerebrospinal fluid as indices of pathological change. *J. Neurol. Neurosurg. Psychiatr.* **44**, 1129–1135

25 Sakimura, K., Kushiya, E., Obinata, M., Odani, S., Takahashi, Y. (1985a). Molecular cloning and the nucleotide sequence of cDNA for neuron-specific enolase messenger RNA of rat brain. *Proc. Natl. Acad. Sci. USA* **82** 7453–7457

26 Sakimura, K., Kushiya, E., Obinata, M., Odani, S., Takabashi, Y. (1985a). Molecular cloning and the nucleotide sequence of cDNA to mRNA for non-

neuronal enolase (αα enolase) of rat brain and liver. *Nucl. Acids Res.* **13**, 4365–4378

27 Scarna, H., Delafosse, B., Steinberg, R., Debilly, G., Mandrand, B., Keller, A., Pujol, J.F. (1982). Neuron specific enolase as a marker of neuronal lesions during various comas in man. *Neurochem. Int.* **4**, 405–411

28 Schmechel, D.E., Brightman, M.W., Marangos, P.J. (1980). Neurons switch from non-neuronal (NNE) enolase to neuronal (NSE) enolase during development. *Brain Res.* **190**, 195–214

29 Schmechel, D.E., Marangos, P.J. (1983). Neuron specific enolase as a marker for differentiation in neurons and neuroendocrine cells. In *Current Methods in Cellular Neurobiology* (J. Barker, J. McKelvey eds.) Wiley, New York **1** pp. 1–62

30 Schmechel, D.E., Marangos, P.J., Brightman, M. (1979). Neuron specific enolase is a marker for peripheral and central neuroendocrine cells. *Nature* **276**, 834–836

31 Seshi, B., Bell, C.E. (1985). Preparation and characterization of monoclonal antibodies to human neuron specific enolase. *Hybridoma* **4**, 13–25

32 Soler Federsppiel, B.S., Cras, P., Ghevens, J., Andries, D., Lowenthal, A. (1987). Human γγ enolase: Two site immunoradiometric assay with a single monoclonal antibody. *J. Neurochem.* **48**, 22–28

33 Whitehead, M.C., Marangos, P.J., Connolly, S.M., Morest, D.K. (1982). Synapse formation is related to the onset of neuron-specific enolase immunoreactivity in the avian auditory and vestibular system. *Dev. Neurosci.* **5**, 298–307

34 Wick, M.R., Bernd, M.D., Scheithauer, W., Kovacs, K. (1983). Neuron specific enolase in neuroendocrine tumors of the thymus, bronchia and skin. *Am. J. Clin. Pathol.* **79**, 703–707

35 Zeltzer, P.M., Marangos, P.J., Parma, A.M., Sather, H., Dalton, A., Siegel, S., Seeger, R.C. (1983). Raised neuron-specific enolase in serum of children with metastatic neuroblastoma. *Lancet 1* **13**, 361–363.

36 Zeltzer, P.M., Marangos, P.J., Sather, H., Evans, A., Siegel, S.,Wong, K.Y., Dalton, A., Seeger, R., Hammond, D. (1985). Prognostic importance of serum neuron specific enolase in local and widespread neuroblastoma. *Adv. Neuroblast. Res.* **175**, 319–329

6 Diagnostic Value of Chromogranin A Measured in the Circulation

Marwan A. Takiyyuddin, Juan A. Barbosa, Ray J. Hsiao, Robert J. Parmer, and *Daniel T. O'Connor*

6.1 Introduction

Chromogranin A, a 48 KD acidic soluble protein, was initially recognized as the major soluble protein in the core of the adrenal medullary catecholamine storage vesicles, the chromaffin vesicles (Fig. 6-1) (Smith and Winkler, 1967; Smith and Kirshner, 1967). Subsequently, chromogranin A immunoreactivity was demonstrated in a wide variety of neuroendocrine secretory vesicles, including those in postganglionic sympathetic nerve endings and in virtually all normal polypeptide hormone producing tissues (Lloyd and Wilson, 1983; O'Connor, 1983; O'Connor et al., 1983a; O'Connor and Frigon, 1984). Among neuroendocrine tissues, the adrenal medulla is the quantitatively major tissue source of chromogranin A (O'Connor, 1983).

Although the primary structure of human chromogranin A has been determined (Konecki et al., 1987; Helman et al., 1988a), its physiologic role remains a matter of conjecture.

In 1984, we developed a radioimmunoassay for human chromogranin A (O'Connor and Bernstein, 1984), based on chromogranin A purified from catecholamine storage vesicles of human pheochromocytoma (O'Connor et al., 1984a). The assay utilized rabbit polyclonal antisera (O'Connor and Bernstein, 1984). We have subsequently developed several assay modifications to optimize assay performance, including rapid immunoprecipitations (O'Connor et al., 1989) and the use of a monoclonal antibody against chromogranin A (Sobol et al., 1986).

Human serum samples are incubated with rabbit anti-human chromogranin A (first antibody), and ^{125}I labeled human chromogranin A, 10,000 cpm/tube. The solution is vortex-mixed and allowed to incubate over night at 4 °C. The

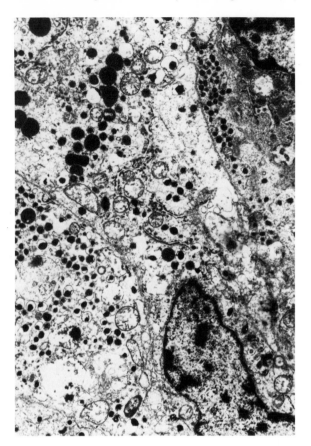

Fig. 6-1 Transmission electron micrograph of a section of human pheochromocytoma. The electron dense core structures are chromaffin vesicles. The freshly obtained operative tissue was fixed, stained, and viewed as previously described (O'Connor and Frigon, 1984). The magnification is 6300 diameters. Reprinted with permission from *Hypertension* (Takiyyuddin et al., 1990b).

Fig. 6-2 In vitro stability of human plasma chromogranin A immunoreactivity. (a) ▶ Immunoreactivity during repeated freeze/thaw cycles (n=6 samples). (b) Immuno-reactivity before and after lyophilization plus water reconstitution (n = 7 samples). (c) Immunoreactivity after prolonged incubation at 37 °C. Reprinted with permission from *Hypertension* (Takiyyuddin et al., 1990b) and *Clinical Chemistry* (O'Connor et al., 1989).

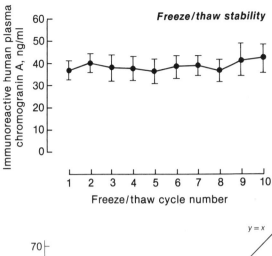

Freeze/thaw stability

Immunoreactive human plasma chromogranin A, ng/ml

Freeze/thaw cycle number

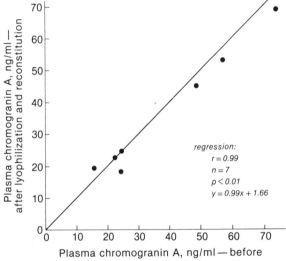

Plasma chromogranin A, ng/ml — after lyophilization and reconstitution

Plasma chromogranin A, ng/ml — before

y = x

regression:
r = 0.99
n = 7
p < 0.01
y = 0.99x + 1.66

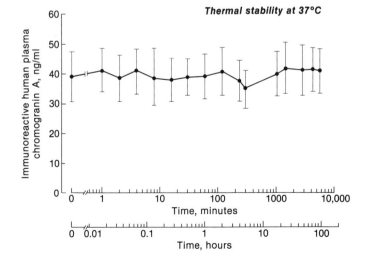

Thermal stability at 37°C

Immunoreactive human plasma chromogranin A, ng/ml

Time, minutes

Time, hours

next day, normal rabbit serum (carrier second antigen) is added, followed by goat anti-rabbit gamma globulin (second antibody), and the solution is vortex-mixed. Aqueous polyethylene glycol is then added to precipitate the immune complexes (leaving the free antigen in solution). The solution is centrifuged at 1500 x g for 20 minutes, the supernate aspirated, and the radioactivity of the pellet is counted (O'Connor et al., 1989).

Circulating chromogranin A immunoreactivity is remarkably stable to a variety of insults, including repeated freezing and thawing (Fig. 6-2a), lyophilization followed by water reconstitution (Fig. 6-2b), and prolonged incubation at 37 °C (Fig. 6-2c).

Measurement of circulating chromogranin A concentration by radioimmunoassay has emerged as a potential probe of exocytotic sympathoadrenal activity in man (O'Connor and Bernstein, 1984), as well as useful diagnostic adjunct in the evaluation of neuroendocrine neoplasia (O'Connor and Deftos, 1986; O'Connor and Deftos, 1987; O'Connor et al., 1989).

In the present chapter, we evaluate the diagnostic value of chromogranin A measured in the circulation under both physiologic and pathologic conditions, and parameters that influence circulating chromogranin A concentrations.

6.2 Chromogranin A Release from the Sympathoadrenal System

In vitro, chromogranin A is coreleased with catecholamines from both adrenal chromaffin cells and sympathetic nerve endings, suggesting exocytosis as the mechanism of catecholamine secretion (Schneider et al., 1967; Smith et al., 1970).

In man, selective provocation of catecholamine secretion from either the adrenal medulla or sympathetic neurons results in corelease of chromogranin A, with consequent increments in plasma chromogranin A. The corelease suggests exocytosis as the mechanism of sympathoadrenal catecholamine secretion *in vivo* (Takiyyuddin et al., 1990a).

Fig. 6-3 illustrates the response of plasma chromogranin A as well as catecholamines to selective adrenomedullary activation by insulin induced hypoglycemia in normal subjects. Following insulin administration, plasma epinephrine rose 14 fold, peaking at 30 minutes, while plasma chromogranin A rose 1.7 fold, peaking at 90 minutes. This temporal dissociation in the appearance of plasma chromogranin A peak concentration (90–120 min) versus catecholamine peak concentration (30–60 min) suggests different modes of transport of chromogranin A and catecholamines from the chromaffin vesicle to the circulation. This is in line with the findings of Carmichael et al. (Carmichael et al., 1990) who observed that, during feline adrenomedullary stimulation, chromogranin A is predominantly transported via a lymphatic route, ultimately involving the thoracic duct, to the circulation, while catecholamines are transported directly into the adrenal vein.

In addition, intense selective activation of sympathetic neurons by vigorous dynamic bicycle exercise (200 watts for 10–12 minutes) induced modest but significant increments (1.2 fold rise) in plasma chromogranin A along with

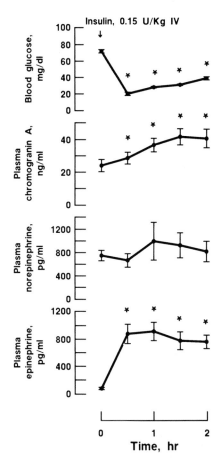

Fig. 6-3 Plasma glucose, catecholamine (norepinephrine) and epinephrine, and chromogranin A responses during insulin induced hypoglycemia. All data points are the mean +/− SEM (n = 12). *p < 0.05. Reprinted with permission from *Circulation* (Takiyyuddin et al., 1990a).

plasma norepinephrine. However, the increments in plasma chromogranin A attained during sympathoneuronal activation are modest compared with increments attained during adrenomedullary activation (1.2 fold versus 1.7 fold). This is consistent with the finding of 97 fold less chromogranin A (ug/g) in sympathetic nerve endings relative to adrenal medulla (Takiyyuddin et al., 1990a).

On the other hand, lesser degrees of adrenal medullary (caffeine ingestion) or sympathoneuronal (standing, smoking, low intensity exercise) activation fail to perturb plasma chromogranin A despite significantly elevating plasma catecholamines (Takiyyuddin et al., 1990a).

Thus, at extremes of sympathoadrenal activity, the mechanism of catecholamine secretion from the adrenal medulla and sympathetic neurons appears to be, at least in part, exocytotic.

6.3 Chromogranin A Release from other Endocrine Sites

Chromogranin A immunoreactivity exhibits a wide endocrine distribution, as revealed by immunohistology, immunoblotting and radioimmunoassay (Lloyd and Wilson, 1983; O'Connor, 1983; O'Connor et al., 1983a; Cohn et al., 1984). Chromogranin A immunoreactivity has been detected in gut enterochromaffin, thyroid calcitonin C, parathyroid chief, adrenal medullary chromaffin, pituitary (anterior > intermediate > posterior), and endocrine pancreatic cells. On the other hand, chromogranin A is not localized to tissues with solely exocrine function or to endocrine tissues producing only nonpeptide hormones (O'Connor, 1983; O'Connor et al., 1983a, O'Connor et al., 1983b; O'Connor et al., 1984b).

In man, stimulation of normal human endocrine secretory cells, other than adrenal medulla (i.e., anterior pituicytes, pancreatic islet cells, gut enteroendocrine cells, parathyroid chief cells, and thyroid parafollicular C-cells) with cell-type selective secretagogues, fails to induce significant increments in plasma chromogranin A despite inducing measurable increments in the concentrations of the resident peptide hormones (O'Connor et al., 1987). Hence, within the neuroendocrine system only stimulation of the adrenal medulla is likely to elevate plasma chromogranin A under physiologic circumstances (Takiyyuddin et al, 1990b). This is consistent with the finding that the adrenal medulla is the quantitatively major neuroendocrine tissue source of chromogranin A (O'Connor, 1983).

6.4 Parameters that Influence Circulating Chromogranin A Concentrations

The source of basal circulating chromogranin A levels in man is still a matter of investigation. Basal circulating chromogranin A displays a significant ultradian or pulsatile rhythm, peaking every 51 minutes (Takiyyuddin et al., 1988). In man, somatostatin infusion suppresses basal circulating chromogranin A levels, and diminishes the magnitude and frequency of pulsatile peaks, without altering plasma catecholamines (Takiyyuddin et al., 1989).

Plasma chromogranin A correlates significantly with serum creatinine concentration ($p < 0.05$)(Hsiao et al., 1990a; O'Connor et al., 1989). In patients with renal insufficiency, plasma chromogranin A is markedly elevated, with the highest values attained in patients with end stage renal disease (Hsiao et al., 1990a; O'Connor et al., 1989). This suggests a major role for the kidneys in the removal of chromogranin A immunoreactivity from circulation. Consequently, knowledge of renal function is essential for proper interpretation of changes in plasma chromogranin A changes.

On the other hand, plasma chromogranin A is elevated to a lesser degree in patients with hepatic disease, precluding a major role for the liver in chromogranin A disposition (O'Connor et al., 1989).

Plasma chromogranin A is modestly but significantly elevated (mean, 1.4 fold) in patients with essential hypertension, suggesting that there may be an excess of exocytotic sympathoadrenal activity in this disorder (O'Connor, 1985; Takiyyuddin et al., 1990b). Suppression of sympathetic outflow with guanabenz, a centrally acting alpha 2 agonist, diminishes blood pressure and suppresses basal circulating chromogranin A levels, suggesting that the plasma chromogranin A basal elevation seen in essential hypertension is, at least in part, neural in origin (O'Connor, 1985). Other antihypertensive therapies, including the angiotensin converting enzyme inhibitor enalapril or the beta-blocker propranolol, lower blood pressure without altering plasma chromogranin A (Takiyyuddin et al., 1990b).

6.5 Chromogranin A as a Marker for Neuroendocrine Neoplasia

Although no functional role for the chromogranins has been conclusively established, potential application of their measurement as diagnostic markers for a variety of neuroendocrine neoplasia has been presented in several reports during recent years.

Chromogranin A immunoreactivity has been demonstrated, by radioimmunoassay and immunoblotting, in dense core secretory vesicles in a variety of polypeptide hormone producing tumors (Lloyd and Wilson, 1983; O'Connor and Deftos, 1986; O'Connor et al., 1984a; Weiler et al., 1988). Chromogranin A may be exocytotically coreleased with the usual resident peptide hormone, suggesting that measurement of plasma chromogranin A may be of value in the diagnosis of such tumors.

Indeed, plasma chromogranin A is elevated in patients with pheochromocytoma, aortic body tumor, carcinoid tumors (with the very highest plasma chromoganin A values observed), pancreatic islet cell tumors, medullary thyroid carcinoma and C-cell hyperplasia, and in a variety of pituitary tumors (O'Connor and Deftos, 1986; Moattari et al., 1989). In these tumors, the diagnostic sensitivity and specificity of plasma chromogranin A measurement has been estimated at 81 % and 100 %, respectively (O'Connor and Deftos, 1987).

In pheochromocytoma, plasma chromogranin A is markedly elevated (10–20 fold), suggesting that catecholamine secretion from the tumor is at least in part exocytotic (O'Connor and Bernstein, 1984; O'Connor and Deftos, 1987; Takiyyuddin et al., 1990b). Elevated plasma chromogranin A (> 112 ng/ml) has a comparable sensitivity (83 %) to, but greater specificity (96 %) than plasma catecholamines in distinguishing pheochromocytoma patients from normals, or other hypertensives (unpublished work). Furthermore, the elevation in plasma chromogranin A correlates with tumor burden and catecholamine production by the tumor, but does not vary with age, tumor location, or sex. In addition, drugs (clonidine, metoprolol, phentolamine, tyramine) that are used in diagnosis or treatment of pheochromocytoma do not perturb plasma chromogranin A, and thus have little effect on the diagnostic accuracy of chromogranin A for pheochromocytoma (unpublished work).

Plasma chromogranin A is elevated in patients with small cell lung carcinoma. Furthermore, plasma chromogranin A correlates with tumor burden, suggesting its potential use as a marker of disease activity (Sobol et al., 1986).

Plasma chromogranin A is also elevated in most patients with Zollinger-Ellison syndrome (ZES) (Stabile et al., 1990). Chromogranin A level correlates with neither gastrin levels, the site and amount of the primary tumor, nor the presence or absence of metastasis (Stabile et al., 1990). Interestingly, plasma chromogranin A levels are reduced (by 66 %) by gastrectomy alone, suggesting the gastric enteroendocrine cells as one source of plasma chromogranin A elevation (Stabile et al., 1990).

We initially found that plasma chromogranin A was elevated in patients with primary hyperparathyroidism, with higher levels attained in patients with parathyroid hyperplasia versus patients with parathyroid adenoma (O'Connor and Deftos, 1986). However, we have recently found that plasma chromogranin A elevation in hyperparathyroidism is largely restricted to patients with familial multiple endocrine neoplasia (MEN I), who also have ZES (Nanes et al., 1989). Furthermore, following parathyroidectomy in patients with either parathyroid adenoma or hyperplasia, plasma chromogranin A is reduced by 54 %, but only in patients with ZES. This suggests that parathyroid gland is not the source of plasma chromogranin A elevation in this condition, but rather that hyperparathyroidism influences chromogranin A release from other tissues, possibly pancreatic islet cells (Nanes et al., 1989) or stomach enteroendocrine cells (Stabile et al., 1990).

Recently, plasma chromogranin A has been shown to be elevated in children with neuroblastoma, and the elevation is greater in widespread or metastatic than in localized disease (Hsiao et al., 1990b). Furthermore, plasma chromogranin A concentration parallels disease stage and can be, either alone or in combination with age or disease stage, an effective prognostic index (Hsiao et al., 1990b).

Plasma chromogranin A, on the other hand, is not secreted by nonendocrine tumors or endocrine tumors not associated with dense core vesicles (O'Connor and Deftos, 1986; O'Connor et al., 1989). Interestingly, chromogranin A mRNA expression has been detected in colon tissues with adenocarcinoma suggesting potential neuroendocrine lineage of some subsets of these tumors (Helman et al, 1988b).

In conclusion, measurement of circulating chromogranin A has emerged as a useful tool for the diagnosis and determination of the extent of neuroendocrine neoplasia. In addition, it has provided further insight into the mechanism of sympathoadrenal catecholamine secretion in man.

Acknowledgements
We appreciate the technical assistance of Ms. Annie Chen and Mrs. Justine Cervenka. This work was supported by the Veterans Administration, the National Institutes of Health, the American Heart Association, and the National Kidney Foundation.

6.6 References

1 Carmichael, S.W., Stoddard, S.L., O'Connor, D.T., Yaksh, T.L. and Tyce, G.M. (1990). The secretion of catecholamines, chromogranin A, and neuropeptide Y from the adrenal medulla of the cat via the adrenolumbar vein and the thoracic duct: different anatomic routes based on size. *Neuroscience* **34**, 433–440

2 Cohn, D.V., Elting, J.J., Frick, M. and Elde, R. (1984). Selective localization of the parathyroid secretory protein I/adrenal medulla chromogranin A protein family in a wide variety of endocrine cells of the rat. *Endocrinology* **114**, 1963–74

3 Helman, L.J., Ahn, T.G., Levine, M.A., Allison, A., Cohen, P.S., Cooper, M.J., Cohn, D.V. and Israel, M.A. (1988a). Molecular cloning and primary structure of human chromogranin A (secretory protein I) cDNA. *J. Biol. Chem.* **263**, 11559–11563

4 Helman, L.J., Gazdar, A.F., Park, J.G., Cohen, P.S., Cotelingam, J.D. and Israel, M.A. (1988b). Chromogranin A expression in normal and malignant human tissues. *J. Clin. Invest.* **82**, 686–690

5 Hsiao, R. J., Mezger, M. S., and O'Connor, D. T. (1990a) Chromogranin A in uremia: progressive retention of immunoreactive fragments. Kidney International **37**, 955–964

6 Hsiao, R.J., Seeger, R.C., Yu, A. and O'Connor, D.T. (1990b). Chromogranin A in children with neuroblastoma: plasma concentration parallels disease stage and predicts survival. *J. Clin. Invest.* **85**, 1555–1559

7 Konecki, D.S, Benedum, U.M., Gerdes, H.H. and Huttner, W.B. (1987). The primary structure of human chromogranin A and pancreastatin. *J. Biol. Chem.* **262**, 17026–17030

8 Lloyd, R.V. and Wilson, B.S. (1983). Specific endocrine tissue marker defined by a monoclonal antibody. *Science* **222**, 628–630.

9 Moattari, A.R., Deftos, L.J. and Vivnik, A.I. (1989). Effects of sandostatin on plasma chromogranin-A levels in neuroendocrine tumors. *J. Clin. End. Metab.* **69**, 902–905

10 Nanes, M.S, O'Connor, D.T. and Marx, S.J. (1989). Plasma chromogranin A in primary hyperparathyroidism. *J. Clin. End. Metab.* **69**, 950–955

11 O'Connor, D.T. (1983). Chromogranin A: widespread immunoreactivity in polypeptide hormone producing tissues and in serum. *Regul. Peptides* **6**, 263–280

12 O'Connor, D.T., Burton, D.W. and Deftos, L.J. (1983a). Chromogranin A: immunohistology reveals its universal occurrence in normal polypeptide hormone producing endocrine glands. *Life Sci.* **33**, 1657–1663

13 O'Connor, D.T., Burton, D.G. and Deftos, L.J. (1983b). Immunoreactive chromogranin A in diverse polypeptide hormone producing tumors and normal endocrine tissues. *J. Clin. End. Metab.* **57**, 1084–1086

14 O'Connor, D.T. and Frigon, R.P. (1984). Chromogranin A, the major catecholamine storage vesicle soluble protein. *J. Biol. Chem.* **259**, 3237–3247

15 O'Connor, D.T. and Bernstein, K.N. (1984). Radioimmunoassay of chromogranin A in plasma as a measure of exocytotic sympathoadrenal activity in normal subjects and in patients with pheochromocytoma. *New Engl. J. Med.* **311**, 764–770

16 O'Connor, D.T., Frigon, R.P. and Sokoloff, R.L. (1984a). Human chromogranin A: purification and characterization from catecholamine storage vesicles of pheochromocytoma. *Hypertension* **6**, 2–12

17 O'Connor, D.T., Parmer, R.J. and Deftos, L.J. (1984b). Chromogranin A: studies in the endocrine system. *Trans. Assoc. Am. Physicians* **97**, 242–250

18 O'Connor, D.T., Pandian, M.R., Carlton, E., Cervenka, J.H. and Hsiao, R.J. (1989). Rapid radioimmunoassay of circulating chromogranin A: in vitro stability, exploration of the neuroendocrine character of neoplasia, and assessment of the effects of organ failure. *Clin. Chem.* **35**, 1631–1637

19 O'Connor, D.T. and Deftos, L.J. (1986). Secretion of chromogranin A by peptide producing endocrine neoplasms. *New Engl. J. Med.* **314**, 1145–1151

20 O'Connor DT. and Deftos L.J. (1987). How sensitive and specific is measurement of plasma chromogranin A for the diagnosis of neuroendocrine neoplasia? *Ann. N.Y. Acad. Sci.* **493**, 379–386

21 O'Connor, D.T., Pandian, M.R., Cervenka, J.H., Mezger, M. and Parmer, R.J. (1987). What is the source and disposition of chromogranin A in normal human plasma? *Clin. Res.* **35**, 605A

22 O'Connor, D.T. (1985). Plasma chromogranin A: initial studies in human hypertension. *Hypertension* **7**, I76-I79

23 Schneider, F.H., Smith, A.D. and Winkler, H. (1967). Secretion from the adrenal medulla: biochemical evidence of exocytosis. *Br. J. Pharmacol. Chemother.* **31**, 94–104

24 Smith, A.D. and Winkler, H. (1967). Purification and properties of an acidic protein from chromaffin granules of bovine adrenal medulla. *Biochem. J.* **103**, 483–492

25 Smith, W.J. and Kirshner, N. (1967). A specific soluble protein from the catecholamine storage vesicles of bovine adrenal medulla. I. Purification and characterization. *Molec. Pharmacol.* **3**, 52–62

26 Smith, A.D., De Potter, W.P., Moerman, E.H. and De Schaepdryver, A.P. (1970). Release of dopamine-beta-hydroxylase and chromogranin A upon stimulation of the sympathetic nerve. *Tissue & Cell* **2**, 547–568

27 Sobol, R.E., O'Connor, D.T., Addison, J., Suchocki, K., Royston, I. and Deftos, L.J. (1986). Elevated serum chromogranin A concentrations in small-cell lung carcinoma. *Ann. Int. Med.* **105**, 698–700

28 Stabile, B.E., Howard, T. J., Passaro, E. and O'Connor D.T. (1988). Source of plasma chromogranin A elevation in gastrinoma patients Arch. Surg. **125**, 451–453

29 Takiyyuddin, M.A., Cervenka, J.H., Sullivan, P.A., Pandian, M.R., Parmer, R.J., Barbosa, J.A. and O'Connor, D.T. (1990a). Is physiologic sympathoadrenal catecholamine release exocytotic in man? *Circulation* **81**, 185–195

30 Takiyyuddin, M.A., Cervenka, J. and O'Connor, D.T. (1988). Pulsatile release of plasma chromogranin A in man. *Am. J. Hypertension* **1**, 43A

31 Takiyyuddin, M.A., Cervenka, J., Baron, A.D. and O'Connor, D.T. (1989). Somatostatin infusion suppresses the pulsatile release of chromogranin A in man. *Clin. Res.* **37**, 461A

32 Takiyyuddin, M.A., Cervenka, J.H., Hsiao, R.J., Barbosa, J.A., Parmer, R.J. and O'Connor, D.T. (1990b). Chromogranin A: Storage and release in Hypertension. *Hypertension* **15**, 237–246

33 Weiler, R., Fischer-Colbrie, R., Schmid, K.W., Feichtinger, H., Bussolati, G., Grimelius, L., Krisch, K., Kerl, H., O'Connor, D.T. and Winkler, H. (1988). Immunological studies on the occurrence and properties of chromogranin A and B and secretogranin II in endocrine tumors. *Am. J. Surg. Pathol.* **12**, 877–884

7 Markers for Neural and Endocrine Cells in Pathology

Philipp U. Heitz, Jürgen Roth, Christian Zuber and *Paul Komminoth*

7.1 Introduction

The use of markers for the phenotypic characterization of cells has yielded a wealth of information on number, types, distribution and function of normal and tumorous neuroendocrine cells. This knowledge enables the pathologist to analyze pathological phenomena of neuroendocrine cells and tissues, i.e. hyperplasia and tumors with more precision than ever. Markers are now extensively used in diagnostic and experimental pathology.

In *diagnostic pathology* the pathologist is very often confronted with the question of whether a given patient suffers from a tumor or from a different type of lesion. In presence of a tumor the next diagnostic steps are 1) to define the phenotype of the tumor, 2) to define the biological behavior, 3) to decide if the tumor is a primary or a metastasis and to localize the primary, and 4) to demonstrate the synthesis of secretory products.

Histology, using conventional tinctorial stains, is still a very important technique in the diagnosis of lesions. However additional techniques, i.e. special tinctorial stains, electron microscopy, immunocytochemistry, *in situ* hybridization and polymerase chain reactions are becoming increasingly important. Some of these techniques are at present widely used to define the phenotype, grade of differentation, biological behavior, and heterogeneity of tumor cells.

All these techniques can be useful in the biological analysis and diagnosis of lesions, especially of tumors, if carefully selected and used properly. At present immunocytochemistry is most widely used for localizing markers because of the possibility of establishing correlations with biochemically demonstrable serum markers, and the monitoring of the course of disease and/or the effect of therapy. It must however be born in mind that immunocytochemical techniques are merely able to localize epitopes on posttranslational products. Per se these methods allow neither the definition of the synthetic activity of the cell, nor the ability of secretion of a given substance by the cell. In order to provide functionally significant results, immunocytochemical methods must be combined with investigations into the state and activity of receptors on the cell membrane, *in situ* hybridization and/or polymerase chain reaction to define the biosynthetic activity of the cell, and with assays determining the serum concentration of a given secretory product. Finally, the biochemistry and site of the conversion of pre/prosecretory products into the final secretory product must be elucidated (Roth et al., 1989).

In medicine the combined findings of 1) the symptoms presented by the patient, 2) one or several posttranslational products localized in inflammatory or, more important, in tumor cells, and 3) an elevated serum concentration of these products, is often sufficient for establishing a precise diagnosis. The application of methods determining these phenomena has revolutionized the diagnosis of tumors, in particular the diagnosis of lymphomas and neuroendocrine tumors.

In this short overview some of the more important general markers of the neuroendocrine phenotype of normal and tumor cells (so called broad-spectrum markers) are discussed. It is beyond the scope of this short review to discuss the numerous markers, i.e. hormones, considered specific for the phenotype of the various neuroendocrine cells.

7.2 Neuroendocrine Tumors

The term neuroendocrine is used here to define the secretory products of the cells and tumors rather than their nature and embryological derivation. Neuroendocrine tumors occur in organs known to secrete neuroendocrine messengers and/or to contain cells displaying characteristics or activity typical of the neuroendocrine phenotype (Heitz, 1987).

Individual neuroendocrine tumor types are uncommon but neuroendocrine tumors as a group are common.

There are several groups of markers characteristic of the neuroendocrine phenotype, including 1) the presence of voltage dependent Na^+ and/or Ca^{++} channels in the cell membrane, 2) receptors for specific ligands, i.e. nerve growth factor, 3) cytoskeletal proteins, 4) soluble proteins like neuron-specific enolase and protein gene product (PGP) 9.5, 5) granule matrix constituents, i.e. chromogranins and epitope Leu-7, 6) secretory vesicle membrane constituents, i.e. cytochrome B 561 or synaptophysin, and 7) amine biosynthetic enzymes. In a near future additional markers, i.e. NCAM (Neural cell adhesion molecule), may provide new insights into the biology of neuroendocrine cells.

Markers for cytoskeletal proteins, neuron-specific enolase, chromogranins, Leu-7, and synaptophysin are currently used for diagnosis and precise classification of tumors. The localization and chemistry of NCAM or NCAM polysialic acid will probably soon be clarified in the various neuroendocrine cells and tumors. It is a promising additional marker of various types of neuroendocrine tumors (see below).

Histology, using conventional or special tinctorial stains, is a prerequisite for the diagnosis of tumors, including neuroendocrine tumors. The morphology of the tumors, and metastases thereof, is often characteristic, at least in highly differentiated tumors. This picture becomes blurred in tumors of a lesser degree of phenotypic differentiation, especially in malignant tumors. Moreover, on the basis of conventional histology it is not possible to reach a precise functionally significant diagnosis, because by morphology alone no conclusion can be drawn as to the production of hormones by a given cell. In this situation the use of markers characteristic for the neuroendocrine phenotype of cells may be helpful in reaching a diagnosis, or at least a differential diagnosis.

The tumors often cause characteristic symptoms, or well defined syndromes by inappropriate hormone secretion. The concentration in the patient's serum of many secretory products can be determined. The pathologist must therefore provide a specific and functionally significant diagnosis. The final evidence for the production and secretion of a given substance can be provided only by the combined effort of the clinician, biochemist, and morphologist. This evidence is of paramount importance for the patient because some tumors are of low grade malignancy, and can therefore be treated successfully, using drugs with a specific action directed against specific types of tumor cells and against effects of secretory products on their target cells. Even in the presence of metastases the prognosis can be much better for a patient suffering from a neuroendocrine tumor than from other tumor types.

7.3 Technical Prerequisites

The quality of appropriate tissue processing, including snap freezing or appropriate fixation, is of utmost importance for the successful use of markers. It cannot be overemphasized that routine fixation in buffered formaldehyde is appropriate or at least sufficient for many reactions. Deparaffinized sections of formaldehyde-fixed tissue can be briefly dipped into Bouin's fluid for obtaining optimal results with some antibodies. A number of antigens withstand low concentrations of glutaraldehyde (0.1 to 1 %) mixed with formaldehyde, or even postfixation with osmium tetroxide. Snap freezing is necessary for biochemical analysis of the tumor tissue and often for reactions using *in situ* hybridization for visualizing mRNA. However, routinely fixed and paraffin-embedded tissue has been shown to be suited for successful localization of mRNA of NCAM (Roth et al., 1988a). On the other hand, the polymerase chain reaction (PCR) is of great promise because of the possible millionfold *in situ* synthesis of a known primer even on conventionally fixed and stained tissue.

The use of purified and specific antibodies and/or probes is essential for obtaining reliable results. The importance of extensive testing for optimal reaction conditions, and the demonstration of the specificity of the reaction with every individual antibody or probe cannot be overstressed.

7.4 Significance of Neuroendocrine Markers

The importance of broad-spectrum markers is to provide the pathologist with a clue to the neuroendocrine phenotype of a tumor, if not previously revealed by conventional histology. Using reactions for a range of markers, many tumors were shown to be of the neuroendocrine phenotype, or at least to contain a variable number of cells bearing the neuroendocrine phenotype admixed with other tumor cells.

The significance of the general markers is their independence of hormone production by tumor cells. They are therefore very valuable in the diagnosis of tumors not producing or secreting specific messengers. As described below, the general neuroendocrine markers can be used in a *second step* of a diagnostic flow chart. The *first step* can consist of the general characterization of a tumor using cytokeratins, vimentin and neurofilament proteins. In the *third step* the presence of specific cell products, i.e. peptide hormones can be visualized.

In the interpretation of the reactions it must be born in mind that there are no absolute criteria for defining the phenotype of a given tumor cell on the basis of the combined occurrence of various types of cytoskeletal proteins. On the other hand there are tumors, not displaying further neuroendocrine characteristics, which yield reactions with one or several neuroendocrine markers. *The differences obtained in the reactions are therefore quantitative rather than qualitative* although many tumors follow some general rules.

No strict rules can be established for the use of a certain number or set of reactions. As mentioned above it must be the morphology which guides the pathologist in the careful selection of an appropriate set of reactions to be carried out.

7.5 Diagnostic Algorithm

In the following three possible steps of a diagnostic flow chart are discussed. The last step, i.e. the visualization of specific cell markers, is discussed only briefly because it is beyond the scope of this short review.

7.5.1 First Step: General Markers

A possible first step in characterizing a tumor is to use antibodies directed against epitopes of cytoskeletal proteins. Many tumor cells of the neuroendocrine phenotype contain cytokeratins, especially cytokeratins number 8 and 18 (Höfler et al., 1986). Antibodies directed against selected cytokeratins or directed against an epitope common to many or all presently known cytokeratins may be used (von Overbeck et al., 1985; Franke et al., 1987). The reaction pattern of the cells may be pancytoplasmic, cell membrane associated or paranuclear "dot-like". The latter pattern is characteristic, although by no means specific, for growth hormone producing pituitary adenomas and neuroendocrine carcinomas of the skin (Merkel cell tumors).

It is striking that the cells of tumors of the adrenal medulla and paragangliomas, steroid hormone producing tumors of the adrenal cortex, testis and ovary, and melanocytes (normal or tumorous) contain only very rarely cytokeratins. Melanocytes and cells of pheochromocytomas or paragangliomas often contain vimentin. The massive vimentin content of endothelial and mesenchymal, and so-called sustentacular cells may produce a very characteristic alveolar like pattern in pheochromocytomas and similar tumors.

It is very interesting that combinations of various cytoskeletal proteins may occur in neuroendocrine tumor cells. Neurofilament proteins may occur in combination with cytokeratins in bronchial and intestinal carcinoid tumors and in bronchial neuroendocrine carcinomas, while vimentin and neurofilaments often occur in the same tumor cells in pheochromocytomas or melanomas. In the latter there is moreover a characteristic combination of vimentin with the cytoplasmic protein S-100. In thyroid tumors, especially in papillary thyroid carcinomas, cells containing either cytokeratin or vimentin often occur combined in the same tumor.

7.5.2 Second Step: General or Broad-Spectrum Neuroendocrine Markers

The second step of a possible diagnostic algorithm comprises general or broad spectrum markers of the neuroendocrine phenotype, i.e. neuron-specific enolase, synaptophysin, PGP 9.5, chromogranins, Leu-7, and NCAM as well as NCAM polysialic acid.

Neuron-Specific Enolase (NSE) and Protein Gene Product (PGP) 9.5

This cytoplasmic enzyme, the chemistry of which is discussed in chapter 5 of this volume, has been used extensively and successfully in morphological analysis (Tapia et al., 1981; Dhillon et al., 1982a, b; Oskam et al., 1985; Iwase et al., 1986; Osborne et al., 1986), and as a serum marker for the characterization of neuroendocrine tumors (Carney et al., 1982; Prinz and

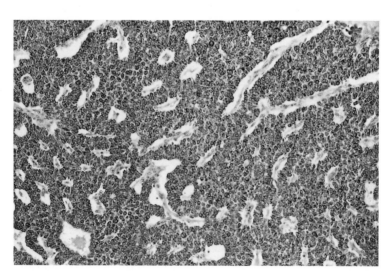

Fig. 7-1 Strong cytoplasmic immunocytochemical reaction for synaptophysin in an insulinoma. Formaldehyde-fixed, paraffin-embedded tissue. Avidin-Biotin-Complex technique, 125×.

Marangos, 1982; Zeltzer et al., 1983). In our experience NSE is a useful marker, provided the tissue is rapidly and thoroughly fixed and a good antibody directed aginast the gamma subunit (see chapter 5) is used. Nonspecific reactions often occur in necrotic or autolytic areas. There is no detectable reaction for NSE in tumors of the thyroid (except medullary carcinoma) and in steroid producing tumors. On the other hand it must be born in mind that so-called NSE, i.e. the gamma-gamma dimer of enolase, may occur in tumors not displaying other neuroendocrine characteristics. The occurrence of NSE has been repeatedly described in solid and cystic neoplasms of the pancreas, a tumor which may be taken for an endocrine neoplasm on the basis of morphology alone. However, in our experience all reactions for other neuroendocrine markers are consistently negative in this tumor (Heitz et al., 1989).

PGP 9.5 is a soluble protein originally isolated from neural tissue (Thompson et al., 1983). Its function is unknown. It has a wide distribution in nerves (Gulbenkian et al., 1987) and occurs in neuroendocrine tumors (Rode et al., 1985). In our experience the marker is not as reliable as others in tumors.

Synaptophysin

This specific component of membranes of presynaptic vesicles apparently occurs in virtually all neurons (see chapter 2) and in at least the majority of normal and tumorous neuroendocrine cells (Fig. 7-1). Its general use has been hampered for some time because good and reliable and also specific antibodies were not commercially available. On the other hand the tissue must be very well preserved, i.e. freshly frozen or well fixed in formaldehyde and embedded in paraffin (Gould et al., 1986; 1987; Wiedenmann et al., 1986). Under these conditions synaptophysin is a very reliable marker for almost all neuroendocrine tumor cells.

Chromogranins

As described in a preceeding chapter of this volume, chromogranins are a group of acidic monomeric proteins of various sizes, which form a major part of the soluble proteins in secretory granules of many endocrine cells (Wilson and Lloyd, 1984). Using specific and sensitive monoclonal antibodies chromogranins have been localized in cells and tumors arising in the adrenal medulla, endocrine pancreas, parathyroid glands, anterior pituitary gland and in the thyroid (C-cells and medullary carcinomas). Focal staining may be observed in

Fig. 7-2 Neuroendocrine cells in the mucosa of a human appendix displaying a strong immunocytochemical reaction for chromogranin A., 400×. Differential interference contrast optics.

Merkel cell tumors, neuroblastoma, and small cell carcinoma of the lung. On the other hand there is in general no chromogranin found in melanomas and naevi. Chromogranins could be localized in the core of cytoplasmic secretory granules by immunoelectron microscopy (Wilson and Lloyd, 1984; Ehrhart et al., 1986; Roth and Heitz, 1989). In addition, chromogranin A was shown to be secreted by at least some of the afore-mentioned tumors (see chapter 3 and 6). The reaction for chromogranins is a useful adjunct in the diagnosis of neuroendocrine tumors yielding good results on formaldehyde-fixed and paraffin-embedded tissue (Fig. 7-2). A condition sine qua non is the presence of a sufficient number of secretory granules containing chromogranin for detection by the techniques presently used. It is therefore obvious that chromogranin is not a reliable marker for tumor cells of a low grade of differentiation, not producing granules. On the other hand chromogranin may be present in tumor cells which can not be shown to produce hormonal products.

Leu-7

Anti-lymphocyte antibody Leu-7 (HNK-1) directed against a carbohydrate epitope present on the cell surface of **h**uman **n**atural **k**iller lymphocytes (Abo and Balch, 1981, Abo et al., 1982), on **myelin-associated glycoprotein** (MAG) (McGarry et al., 1983, Tucker et al., 1984) and on certain cell adhesion molecules yields an intense reaction with normal and neoplastic adrenal medullary cells (Caillaud et al., 1984). The gene coding for the epitope is localized on human chromosome 11. Immunocytochemical studies and immunoblot analysis showed that this antibody apparently reacts with a carbohydrate moiety of a protein (molecular weight 75 000 daltons) localized in the matrix of chromaffin granules. The antibody was shown to react with a small percentage of normal cells of the pancreatic islets, the anterior pituitary, and endocrine cells of the gastrointestinal tract (Tischler et al., 1986). We found a reaction in tumor cells of formaldehyde-fixed, paraffin-embedded tissue of medullary thyroid carcinomas, pheochromocytomas, paragangliomas, endocrine pancreatic tumors, and carcinoids of the stomach, duodenum, ileum and rectum (unpublished results). It has been suggested that Leu-7 immunoreactivity may be a marker of specific subsets of secretory granules. However, the Leu-7 immunoreaction is also obtained on the cell surface of small cell lung cancer (Bunn et al., 1985). This may be due to the presence of Leu-7 (HNK-1) on subsets of NCAM or related molecules (see chapter 4).

Neural Cell Adhesion Molecule (NCAM)

NCAM is a general calcium-independent cell adhesion molecule which serves as a homophilic ligand mediating cell-cell adhesion and affecting a wide variety of different cellular events (Edelman, 1986; Rutishauser and Jessell, 1988; Rutishauser et al., 1988). The NCAM polypeptides exist in various isoforms, the 180 kDa, 140 kDa and 120 kDa polypeptides being most commonly found (Nybroe et al., 1988). The biochemistry, molecular biology and cellular distribution of NCAM is discussed in chapter 4. The homophilic binding properties of NCAM are modulated by differential expression of homopolymers of alpha 2,8-linked N-acetylneuraminic acid, the so-called polysialic acid: the highly sialylated form of NCAM (NCAM-H) exhibits reduced adhesive properties whereas the less sialylated NCAM (NCAM-L) promotes adhesion. Although NCAM is transiently expressed in various early embryonic structures, the H- or L- form of NCAM is found predominantly in neurons and glial cells (Nybroe et al., 1988), in striated muscle (Moore and Walsh, 1985), as well as in the adrenal medulla (Langley and Aunis, 1984), pheochromocytoma cells PC 12 (Margolis and Margolis, 1983), neuroblastoma cells (Livingstone et al., 1988), anterior pituitary (Langley et al., 1987), pancreatic islets (Langley et al., 1989) developing kidney (Roth et al., 1987, 1988a), and Wilms tumor (Roth et al., 1988 a, b). Various monoclonal and polyclonal NCAM antibodies have been demonstrated as useful for immunohistochemistry on native frozen sections or paraffin sections of formaldehyde-fixed tissue. A mouse monoclonal IgG2a antibody recognizing polysialic acid typically found on NCAM-H (Frosch et al., 1985) has been shown to yield excellent results on paraffin sections of routinely formaldehyde-fixed and paraffin-embedded tissues.

It is at present obvious that NCAM-L and NCAM-H can be only auxiliary tools in the identification of neuroendocrine cells. The situation is complex and far from being investigated in detail. In neural tumors the NCAM-H may be prevailing (Livingstone et al., 1988), and in our unpublished studies we found this to be also the case for pituitary adenomas, small cell lung carcinoma, medullary thyroid carcinoma and immature teratomas. In all these instances the monoclonal anti-polysialic acid antibody gave unambiguous results and in the case of the immature teratomas proved to be a superior reagent as compared to other antibodies such as anti-NSE, anti-chromogranin and anti-S100. In our experience, polysialic acid is an excellent marker, provided that structural preservation of the tissue is good. In necrotic or autolytic areas usually no reaction for polysialic acid is detected. This is in accordance with its well known structural fragility. Endocrine cells of the pancreas and insulinomas appear to express only the NCAM-L. Consequently, the polysialic acid antibody yields no reaction, whereas positive immunostaining with NCAM antibodies can be observed. It is not yet clear, if the differential expression of NCAM, NCAM-H, or NCAM-L is related to the grade of differentiation and/or malignancy. Since NCAM is not a component of endocrine secretory granules one may expect to find it in tumors not producing hormones.

7.5.3 Third Step: Specific Markers

Peptides, glycoproteins, biogenic amines, catecholamines and, to a limited extent, steroid hormones may be used as specific markers in a third step of an algorithm designed for the precise typing of neuroendocrine tumor cells. The immunocytochemical localization of these products, of chromogranins and Leu-7 is based on the fact that the intracellular pathway of the substance is regulated as well as directed. In a steady state of the cell metabolism, a post-Golgi pool of secretory products, precursors thereof, and substances contained in the matrix of secretory granules exists. The secretory products can be stored for several hours in granules (Kelly, 1985; Lauffer et al., 1985; Wickner and Lodish, 1985), and therefore a high concentration of secretory products and other substances contained in the granular matrix is present in this post-Golgi compartment.

The extensive use of these markers has provided the cell biologist and the pathologist with new insights into the process of hormone biosynthesis, the secretion and the regulation thereof, and into the phenotypic heterogeneity of tumor cells.

In addition the use of these markers has revolutionized the classification and diagnosis of neuroendocrine tumors, i.e. tumors of the pituitary gland (Heitz et al., 1987), the lung, the gastrointestinal tract and the endocrine pancreas (Klöppel and Heitz, 1988).

The combined use of clinical, biochemical, and functional-morphological analysis of hormone secretion, biosynthesis at the transcriptional, translational and post-translational levels, intracellular transport and conversion of precursors into the final secretory products, and of receptors will undoubtedly provide further exciting findings in the future.

7.6 References

1 Abo, T., Balch, C. M. (1981). A differentiation antigen of human NK and K cells identified by a monoclonal antibody (HNK-1). *J. Immunol.* **127,** 1024–1029

2 Abo, T., Cooper, M. D., Balch, C. M. J. (1982). Postnatal expansion of the natural killer and killer cell population in humans identified by the monoclonal HNK-1 antibody. *J. Exp. Med.* **155,** 321–326

3 Bunn, P. A., Linnoila, I., Minna, J. D., Carney, D., Gazdar, A. F. (1985). Small cell lung cancer, endocrine cells of the fetal bronchus and other neuroendocrine cells express the Leu-7 antigenic determinant present on natural killer-cells. *Blood* **65,** 764–768

4 Caillaud, J. M., Benjelloun, S., Bosq, J., Braham, K., Lipinski, M. (1984). HNK-1 defined antigen detected in paraffin-embedded neuroectoderm tumors and those derived from cells of the amine precursor uptake and decarboxylation system. *Cancer Research* **44,** 4432–4439

5 Carney, D. N., Ihde, D. C., Cohen, M. H., Marangos, P. J., Bunn, P. A. Jr, Minna, J. D., Gazdar, A. F. (1982). Serum neuron-specific enolase: A marker for disease extent and response to therapy of small-cell lung cancer. *Lancet* **I,** 583–585

6 Dhillon, A. P., Rode, J. (1982a). Patterns of staining for neurone specific enolase in benign and malignant melanocytic lesions of the skin. *Diagnostic Histopathology* **5,** 169–174

7 Dhillon, A. P., Rode, J., Leathem, A. (1982b). Neurone specific enolase: an aid to the diagnosis of melanoma and neuroblastoma. *Histopathology* **6,** 81–92

8 Edelmann, G. M. (1986). Cell adhesion molecules in the regulation of animal form and tissue pattern. *Ann. Rev. Cell Biol.* **2,** 81–116

9 Ehrhart, M., Grube, D., Bader, M. F., Aunis, D., Gratzl, M. (1986). Chromogranin A in the pancreatic islet: Cellular and subcellular distribution *J. Histochem. Cytochem.* **34,** 1673–1682

10 Franke, W. W., Winter, S., von Overbeck, J., Gudat, F., Heitz, Ph. U., Staehli, C. (1987). Identification of the conserved, conformation dependent cytokeratin epitope recognized by monoclonal antibody (lu-5). *Virchows Arch. A Pathol. Anat.* **411,** 137–147

11 Frosch, M., Görgen, I., Boulnois, G. J.,Timmis, K. N., Bitter-Suermann, D. (1985). NZB mouse system for production of monoclonal antibodies to weak bacterial antigens: Isolation of an IgG antibody to the polysaccharide capsules of Escherichia coli K1 and group B meningococci. *Proc. Natl. Acad. Sci.* USA **82**, 1194–1198

12 Gould,V. E., Lee, I.,Wiedenmann, B., Moll, R., Chejfec, G., Franke,W.W. (1986). Synaptophysin: A novel marker for neurons, certain neuroendocrine cells and their neoplasms. *Hum. Pathol.* **17,** 979–983

13 Gould,V. E.,Wiedenmann, B., Lee, I., Schwechheimer, K., Dockhorn-Dworniczak, B., Radosevich, J. A., Moll, R., Franke,W.W. (1987). Synaptophysin expression in neuroendocrine neoplasms as determined by immunocytochemistry. *Am. J. Pathol.* **126,** 243–257

14 Gulbenkian, S.,Wharton, J., Polak, J. M. (1987).The visualisation of cardiovascular innervation in the guinea pig using an antiserum to protein gene product 9.5. *J. Auton. Nerv. Syst.* **18,** 235–247

15 Heitz, Ph. U. (1987). Neuroendocrine tumor markers. *Curr. Top. Pathol.* **77,** 279–306

16 Heitz, Ph. U., Landolt, A. M., Zenklusen, H. R., Kasper, M., Reubi, J. C., Oberholzer, M., Roth, J. (1987). Immunocytochemistry of pituitary tumors. *J. Histochem. Cytochem.* **35,** 1005–1011

17 Heitz, Ph., Stolte, M., and Kloeppel (1989) Large pancreatic tumor in a young female. *Ultrastruct. Pathol.* **13**, 589–592

18 Höfler, H., Denk, H., Lackinger, E., Helleis, G., Polak, J. M., Heitz, Ph. U. (1986). Immunocytochemical demonstration of intermediate filament cytoskeleton proteins in human endocrine tissues and (neuro-)endocrine tumors. *Virchows Arch. A Pathol. Anat.* **409,** 609–626

19 Iwase, K., Nagasaka, A., Kato, K., Ohtani, S., Nagatsu, I., Ohyama,T., Nakai, A., Aono,T., Nakagawa, H., Shinoda, S., Inagaki, M.,Tei,T., Miyakawa, S., Kawase, K., Miura, K., Nakamura,T., Kanno,T., Kuwãyama, A., Kageyama, N. (1986). Enolase subunits in patients with neuroendocrine tumors. *Clin. Endocrinol. Metab.* **63,** 94–101

20 Kelly, B. (1985). Pathways of protein secretion in eukaryotes. Science **230,** 25–32

21 Kloeppel, G., Heitz, Ph. U. (1988). Pancreatic endocrine tumors. *Pathol. Res. Pract.* **183,** 155–168

22 Langley, O. K., Aunis D. (1984). Ultrastructural immunocytochemical demonstration of D2-protein in adrenal medulla. *Cell Tissue Res.* **238,** 497–502

23 Langley, O. K., Aletsee M. C., Gratzl, M. (1987). Endocrine cells share expression of N-CAM with neurones. *FEBS Lett.* **220,** 108–112

24 Langley, O. K., Aletsee-Ufrecht, M. C., Grant, N. J., Gratzl, M. (1989). Expression of the neural cell adhesion molecule NCAM in endocrine cells *J. Histochem. Cytochem.* **37,** 781–791

25 Lauffer, L., Garcia, P. D., Harkins, R. N., Coussens, L., Ullrich, A.,Walter, P. (1985). Topology of signal recognition particle receptor in endoplasmic reticulum membrane. *Nature* **318,** 334–388

26 Livingstone, B. D., Jacobs, J. L., Glick, M. C.,Troy, F. A. (1988). Extended polysialic acid chains (n > 55). in glycoproteins from human neuroblastoma cells. *J. Biol. Chem.* **263,** 9443–9448

27 Margolis, R. K., Margolis, R. U. (1983). Distribution and characteristics of polysialosyl oligosaccharides in nervous tissue glycoproteins. *Biochem. Biophys. Res. Commun.* **116,** 889–894

28 McGarry, R. C., Helfand, S. L., Quarles, R. H., Roder, J. C. (1983). Recognition of myelin associated glycoprotein by the monoclonal antibody HNK-1. *Nature* **306,** 376–378

29 Moore, S. E., Walsh, F. S. (1985). Specific regulation of N-CAM/D2-CAM cell adhesion molecule during skeletal muscle development. *EMBO J.* **4,** 623–630

30 Nybroe, O., Linnemann, D., Bock, E. (1988). NCAM biosynthesis in brain. *Neurochem. Int.* **12,** 251–262

31 Osborn, M., Dirk, T., Käser, H., Weber, K., Altmannsberger, M. (1986). Immunohistochemical localization of neurofilaments and neuron specific enolase in 29 cases of neuroblastoma. *Am. J. Pathol.* **122,** 433–442

32 Oskam, R., Rijksen, G., Lips, C. J. M., Staal, G. E. J. (1985). Enolase isozymes in differentiated and undifferentiated medullary thyroid carcinomas. *Cancer* **55,** 394–399

33 Prinz, R. A., Marangos, P. J. (1982). Use of neuron-specific enolase as a serum marker for neuroendocrine neoplasms. *Surgery* **92,** 887–889

34 Rode, J., Dhillon, A. P., Doran, J. F. (1985). PGP 9.5 – new marker for human neuroendocrine tumours. *Histopathology* **9,** 147–158

35 Roth, J., Heitz, Ph. U. (1989). Immunolabeling with the Protein A-Gold Technique. An overview. *Ultrastruct. Pathol.* **13,** 467–484

36 Roth, J., Kasper, M., Stamm, B., Haecki, W. H., Storch, M.-J., Madsen, O. D., Kloeppel, G., Heitz, Ph. U. (1989). Localization of proinsulin and insulin in human insulinoma: Preliminary immunohistochemical results. *Virchows Arch. B Cell. Pathol.* **56,** 287–292

37 Roth, J., Taatjes, D. J., Bitter-Suermann, D., Finne, J. (1987). Polysialic acid units are spatially and temporally expressed in developing postnatal rat kidney. *Proc. Natl. Acad. Sci. USA* **84,** 1969–1973

38 Roth, J., Wagner, P., Zuber, C., Weisgerber, C., Heitz, Ph. U., Goridis, C., Bitter-Suermann, D. (1988a). Reexpression of poly (sialic acid) units of the neural cell adhesion molecule in Wilms tumor. *Proc. Natl. Acad. Sci. USA* **85,** 2999–3003.

39 Roth, J., Zuber, C., Wagner, P., Blaha, I., Bitter-Suermann, D., Heitz, Ph. U. (1988b). Presence of the long chain form of polysialic acid of the neural cell adhesion molecule in Wilms' tumor: identification of a cell adhesion molecule as an onco-developmental antigen and implications for tumor histogenesis. *Am. J. Path.* **133,** 227–240

40 Rutishauser, U., Jessell, T. M. (1988). Cell adhesion molecules in vertebrate neural development. *Physiol. Rev.* **68,** 819–857

41 Rutishauser, U., Acheson, A., Hall, A. K., Mann, D. M., Sunshine, J. (1988). The neural cell adhesion molecule (NCAM) as a regulator of cell-cell interactions. *Science* **240,** 53–57

42 Tapia, F. J., Barbosa, A. J. A., Marangos, P. J., Polak, J. M., Bloom, S. R., Dermody, C., Pearse A. G. E. (1981). Neuron-specific enolase is produced by neuroendocrine tumors. Lancet **I,** 808–811

43 Thompson, R. J., Doran, J. F., Jackson, P., Dhillon, A. P., Rode, J. (1983). A new marker for vertebrate neurons and neuroendocrine cells. *Brain Res.* **278,** 224–228

44 Tischler, A. S., Mobtaker, H., Mann, K., Nunnemacher, G., Jason, W. J., Dayal, Y., DeLellis, R., Adelman, L., Wolfe, H. J. (1986). Anti-lymphocyte antibody Leu-7 (HNK-1) recognizes a constituent of neuroendocrine granule matrix. *J. Histochem. Cytochem.* **34,** 1213–1216

45 Tucker, G. C., Aoyama, H., Lipinski, M., Tursz, T., Thiery, J. P. (1984). Identical reactivity of monoclonal antibodies HNK-1 and NC-1: conservation in vertebrates on cells derived from the neural primordium and on some lymphocytes. *Cell. Differ.* **14,** 223–230

46 von Overbeck, J., Stähli, C., Gudat, F., Carmann, H., Lautenschlager, C., Dürrmüller, U., Takacs, B., Miggiano, V., Stähelin, T., Heitz, Ph. U. (1985). Immunocytochemical characterization of an anti-epithelial monoclonal antibody (mAB Lu-5). *Virchows Arch. A Pathol. Anat.* **407,** 1–12

47 Wickner, W. T., Lodish, H. F. (1985). Multiple mechanisms of protein insertion into and across membranes. *Science* **230,** 400–407

48 Wiedenmann, B., Franke, W. W., Kuhn, C., Moll, R., Gould, V. E. (1986). Synaptophysin: A marker protein for neuroendocrine cells and neoplasms. *Proc. Natl. Acad. Sci. USA* **83,** 3500–3504

49 Wilson, B. S., Lloyd, R. V. (1984). Detection of chromogranin in neuroendocrine cells with a monoclonal antibody. *Am. J. Pathol.* **115,** 458–468

50 Zeltzer, P. M., Parma, A. M., Dalton, A., Siegel, S. E., Marangos, P. J., Sather, H., Hammond, D., Seeger, R. C. (1983). Raised neuron-specific enolase in serum of children with metastatic neuroblastoma. *Lancet* **II,** 361–363

Index